Experimental Results From The Design Institute for Physical Poperty Data: Phase Equilibria and Pure Component Properties

Cline Black, editor

Phase Equilibria

J.W. Blasdel
C. Jeffrey Brady
J.R. Cunningham
L.R. Dohmen
Aa. Fredenslund
James R. Freeman
P.C. Gillespie
Jae Youn Kim
S.O. Lundell
D.B. Manley
J.L. Oscarson
J.L. Owens
B.E. Poling
G. Radnai
P. Rasmussen
R.W. Rousseau
W. Vincent Wilding
G.M. Wilson
Loren C. Wilson

Pure Component Properties

T.E. Daubert
V. Goren
J. Jalowka

AIChE Symposium Series

Number 256 — 1987 — Volume 83

Published by

American Institute of Chemical Engineers

345 East 47 Street — New York, New York 10017

Library of Congress Cataloging-in-Publication No. 87-19345

ISBN 0-8169-0413-8

Printed in the United States of America by
Twin Production & Design

FOREWORD

This is the second volume of the AIChE Symposium Series which presents experimental results from projects sponsored by the AIChE Design Institute for Physical Property Data (DIPPR). They were developed under the auspices of the Experimental Mixture Properties Project 805 and the Pure Component Experimental Data Project 821. Results from Project 805 are presented in the first section titled "Phase Equilibria" and those results from Project 821 are found in the second section entitled "Pure Component Properties".

Under Phase Equilibria is found the experimental data measured for thirty nine binary systems of industrial importance or of value in developing correlation methods for extending and predicting such data. In the section Pure Component Properties, measured vapor pressure data are reported for twenty two pure components for which few or no reliable data were available previously.

The members of the Steering Committees for the two projects have contributed by selecting systems for study, coordinating the research and reviewing the data and correlations. The contributions of those listed below are greatly appreciated.

Cline Black, *editor*

MEMBERS OF STEERING COMMITTEES
PROJECT 805

John Adams
M.A. Albright
H.E. Barner
M.S. Benson
Cline Black
Evan Buck
J.D. Chase
C.F. Chueh
K. Claiborne
George Daniels
C.H. Deal
S.H. DeYoung
C.A. Hillhouse
A. Kivnick
A.H. Larsen
S. Matsumiya
A.D. Meyers
H. Ozkardesh
D.A. Palmer
L. Scherck
T.B. Selover
T.T. Shih
L.M. Shipman
N.D. Smith
W.L. Sutor
E.A. Turek
J. Vidal
R.R. Wood
Chang Wu
D. Zudkevitch

David Zudkevitch Chairman Steering Committee 1980-1984
Cline Black Chairman Steering Committee 1985

PROJECT 821

M.W. Abernathy
M.A. Albright
D.S. Arnold
G.W. Bentzen
Robert R. Buch
Evan Buck
J. David Chase
C.F. Chueh
Beryl Edmonds
Bruce A. Feay
W.B. Fisher
Arnold Kivnick
S.C. McHaney
C.D. Murphy
R.D. Myers
M.E. Rusak
W.H. Seaton
Jean Vidal
K.W. Williamson
Robert Wood

Evan Buck Chairman Steering Committee 1982-1985

DIPPR PROJECT 805 SPONSORS

Air Products and Chemicals, Inc.
Allied Corporation
Amoco Chemicals Corporation
ARCO Chemical Company (Oxirane International)
The ChemShare Corporation
E.I. DuPont de Nemours & Co., Inc.
El Paso Products Company
Ethyl Corporation
Exxon Research & Engineering Company
Halcon SD Group, Inc.
Hooker Chemical Company
Institute Francais du Petrole (IFP)
Institute of Gas Technology
Kerr-McGee Chemical Corporation
M. W. Kellogg Company
The Lummus Company
Norsk Hydro a.s.
Olin Chemicals
Pennwalt Corporation
Phillips Petroleum Company
Pullman Kellogg, Inc.
Shell Development Company
Simulation Sciences, Inc.
Standard Oil Company
Tennessee Eastman Company
Texaco, Inc.
Texasgulf, Inc.
Toyo Engineering Corporation
Union Carbide Corporation

DIPPR PROJECT 821 SPONSORS

Air Products and Chemicals, Inc.
Allied Corporation
American Petrofine, Inc.
Amoco Chemicals Corporation
Celanese Corporation
Dow Corning Corporation
El Paso Products Company
Halcon R&D Corporation
Institute Francais du Petrole (IFP)
Kerr-McGee Chemical Corporation
Olin Chemicals
Pennwalt Corporation
Phillips Petroleum Company
Rohm & Haas Company
Shell Development Company
Tennessee Eastman Company
The Institution of Chemical Engineers
Union Carbide Corporation

SUMMARY OF DATA REPORTED IN THE SECOND AIChE SYMPOSIUM SERIES ON EXPERIMENTAL DATA

Year	Project	Investigators	Systems
Phase Equilibria			
1980	805/80B	John L. Oscarson Scott O. Lundell J.R. Cunningham	Carbonyl Sulfide + Methanol Methanol + Pyridine n-Hexane + Vinyl Acetate Vinyl Acetate + Methyl Ethyl Ketone Methyl Ethyl Ketone + Phenol Propylene + 1,3-Butadiene 1-Butene + Methyl Tertiary-butyl Ether Acetaldehyde + 1-Pentene N-Hexane + 0-Cholortoluene Carbon Disulfine – Methanol
1981	805/81A	Jonathon L. Owens C. Jeffrey Brady James R. Freeman W. Vincent Wilding Grant M. Wilson	1,2-Dichloroethane + Vinyl Chloride 1,1,2-Trichloroethane + Vinyl Chloride Acetonitrile + Vinyl Chloride Acetonitrile + Ethylacetylene Acetonitrile + Vinylacetylene Triethylamine + Methyl Ethyl Ketone
1981	805/81C	R.W. Rousseau Jae Youn Kim	Methanol + Ethyl Mercaptan
1983	805/83	W. Vincent Wilding Loren C. Wilson Grant M. Wilson	Benzoic Acid + Benzonitrile Methyl Tertiary-butyl Ether + Propylene Acetic Anhydride + Methyl Iodide Oxide Dimethylformamide + Water Methyl Acrylate + Acrylic Acid N-Methyl Pyrrolidone + Chloroform Pyridine – Furan N-Methyl Pyrrolidone + p-Xylene Dimethylformamide + 1,3,-Butadiene Diethanolamine + Ethylene Glycol
1984	805/84	G. Radnai P. Rasmussen Aa. Fredenslund	Trichorotrifluoro ethane + n-Hexane Propionaldehyde + Methyl Acetate Propionaldehyde + Ethyl Acetate Methyl Acetate + Butyraldehyde Butyraldehyde + Propyl Acetate
1984	805/84	W. Vincent Wilding Loren C. Wilson Grant M. Wilson	Ethyl Acetate + Trimethylamine Tertiary Butyl Acetate + Isobutane Dichloroethane + Hydrogen Chloride Dimethylformamide + 1-Butene
1985	805/85	J.W. Blasdel B.E. Poling D.B. Manley	Acetonitrile + Diethylamine
1985	805/85	L.R. Dohmen J.W. Blasdel B.E. Poling D.B. Manley	Cyclohexanone Oxime + Cyclohexanone
1981 1982	805/81A 805/82A	Paul C. Gillespie W. Vincent Wilding Grant M. Wilson	Ammonia + Water

SUMMARY OF DATA REPORTED IN THE SECOND AIChE SYMPOSIUM SERIES ON EXPERIMENTAL DATA *(Continued)*

Year	Project	Investigators	Systems
Pure Component Properties			
			Compounds
1982-84	821	R. Bartakovits T.E. Daubert J. Jalowka V. Goren	1 Isopropyl Alcohol 2 Phenol 3 n-Dodecyl Benzene 4 p-Ethyltoluene 5 Methyl-t-butyl ether 6 Monoethanolamine 7 Diethanolamine 8 2,4-Toluenediamine 9 Hexamethylenediamine 10 Hexamethyleneimine 11 Diethylene glycol 12 Triethylene glycol 13 Propylene glycol mono-methyl ether acetate 14 Vinyl Acetate 15 Dimethyl succinate 16 Ethyl acrylate 17 Methacrylic acid 18 ϵ-Caprolactam 19 N-Cyclohexylpyrrolidone 20 Dimethyl sulfoxide 21 Trimethoxysilane 22 3,3,3-Trifluropropene

CONTENTS

PHASE EQUILIBRIA FOR TEN BINARY SYSTEMS

J.L. Oscarson, S.O. Lundell and J.R. Cunningham ■ Department of Chemical Engineering,
Brigham Young University, Provo, UT 84602

Vapor-liquid equilibrium data were measured for nine binary systems: carbonyl sulfide + methanol, methanol + pyridine, n-hexane + vinyl acetate, vinyl acetate + methyl ethyl ketone, methyl ethyl ketone + phenol, propylene + 1,3-butadiene, 1-butene + methyl tertiary-butyl ether, acetaldehyde + 1-pentene, and n-hexane + ortho-chlorotoluene. Liquid-liquid equilibrium data were measured for the binary carbon disulfide + methanol. Pressures, liquid mole fractions, x, and vapor mole fractions, y, (second liquid mole fractions in the case of carbon disulfide + methanol) were determined for each binary pair at four temperatures. These values were measured at a minimum of four and usually five compositions at each temperature. The pure component vapor pressures of methanol, ortho-chlorotoluene, 1-butene, methyl tertiary-butyl ether, propylene, 1,3 butadiene, pyridine, vinyl acetate, n-hexane, methyl ethyl ketone, acetaldehyde and 1-pentene were measured as a function of temperature. These measurements were done under contract as part of an effort by the Design Institute for Physical Properties (DIPPR) of the AIChE to obtain and correlate data on systems important to designing industrial processes.

Distillation is the most commonly used separation process in the chemical industry. In the design of distillation equipment, mass and energy balances are dependent on equilibrium relationships between the liquid and vapor phases. While the energy balance influences the cost of the separation, equilibrium relationships usually indicate whether the separation is even possible.

Presently, vapor-liquid equilibrium behavior cannot be successfully predicted and empirical methods must be used. The design engineer usually employs empirical correlating equations which contain adjustable parameters which are determined by the use of data. These equations can only be used with confidence for interpolation and limited extrapolation. The phase equilibria were measured in this project so that the data base for such correlations could be expanded.

The need for data, and correlations, on systems containing components of interest at temperatures, pressures, and compositions which are close to those encountered in separation processes has been recognized. Measurement of these data is part of the undertaking by the Design Institute for Physical Properties (DIPPR) of the American Institute of Chemical Engineers (AIChE), an organization comprised of many major chemical and design companies. The project described herein was commissioned and sponsored by DIPPR.

The ten binary systems measured at BYU's laboratory were:

1. carbon disulfide + methanol
2. carbonyl sulfide + methanol
3. methanol + pyridine
4. n-hexane + vinyl acetate
5. vinyl acetate + methyl ethyl ketone
6. methyl ethyl ketone + phenol
7. propylene + 1,3-butadiene
8. 1-butene + methyl tertiary-butyl ether
9. acetaldehyde + 1-pentene
10. n-hexane + ortho-chlorotoluene

The phenol + methyl ethanol amine system was attempted but the system was so chemically reactive that reproducible results could not be obtained.

EXPERIMENTAL PROCEDURE

Three experimental methods were used to measure the phase equilibria of the systems listed above. The methanol +

Sponsored by the Design Institute for Physical Properties of the American Institute of Chemical Engineers. Project number 805/80B.

carbon disulfide system exhibited liquid-liquid equilibrium and the liquid phase compositions were determined using calorimetry at 273 K and 293 K and by measuring the total pressure and temperature of a system with a known charge composition (PTx measurements) at 313 and 333 K. The vapor-liquid equilibrium compositions of the methanol + carbonyl sulfide system at all temperatures and hexane + ortho-chlorotoluene system at 353 K were determined by making PTx measurements. The phase behavior of the remaining systems was determined by measuring the pressure and temperature of the system and the compositions of both the vapor and the liquid (PTxy measurements).

PTxy Measurements

The PTxy measurements were made in a reflux bubble cap still which was based on a design developed by Grant M. Wilson. A schematic of the still is shown in Figure 1. This still was made from a one liter stainless steel cylinder cut and flanged at the center. A bubble cap is positioned inside so as to capture generated vapor. A 2.54 cm (1 in.) stainless steel pipe extends downward from the bubble cap to the bottom of the cylinder and then about 15 cm below. The portion of the pipe which extends below the cylinder was wrapped with a resistance heater. Two holes, one on each side of the pipe, just inside the bottom of the cylinder, allow the liquid to circulate between the pipe and the rest of the cell. The cell contains windows, positioned at the level of the bubble cap to allow visual inspection of the liquid level in the cell. A copper constantan thermocouple inserted into the thermowell which extends up through the pipe to just below the bottom of the bubble cap was used to monitor the temperature. A vapor sample line allows the removal of vapor sample from the bubble cap and liquid sample line allows the removal of liquid samples from a position in the pipe below the bubble cap. Extending upward from the still is a reflux condenser which condenses the escaping vapors and refluxes them back to the still. The top of the condenser is connected by a 0.3175 cm (1/8 in.) stainless steel tube to a ballast tank and pressure gauge. The assembly below the condenser was insulated with approximately 6 cm of glass wool.

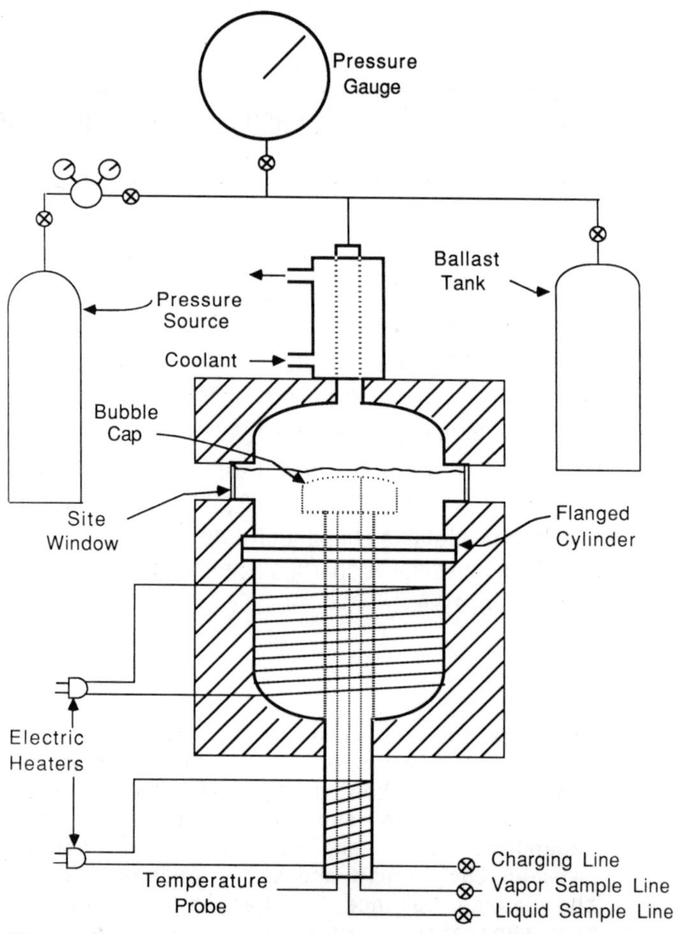

Figure 1. Schematic of reflux bubble cap still used in PTxy measurements.

The still was first flushed with an inert gas, usually nitrogen, and then filled to a level just above the bubble cap with a liquid of the desired composition. The pressure of the inert gas in the cell and ballast tank were increased to a pressure estimated to give the desired temperature. The cooling fluid to the condenser was started and the heaters on the cell were turned on. Once boiling started, the system pressure was adjusted so as to obtain the desired temperature. The ballast tank provided an increased volume, virtually eliminating pressure fluctuations.

As liquid in the inner portion of the still boils, vapor moves upward through the pipe and is captured in the bubble cap. Excess vapor escapes from the bottom of the cap and, together with that generated in the outer portion of the still, moves upward into the condenser, condenses, and returns to the still. The holes in the side of the pipe allow liquid to circulate throughout the still ensuring the liquid is well mixed.

Once the system reached steady state, as evidenced by the temperature no longer changing, samples were withdrawn through the two sample lines and analyzed. The lower heater was turned off while withdrawing a liquid sample to prevent the sample from vaporizing in the line.

The reflux bubble cap still has several advantages as follows:

1. Equilibrium can be achieved without agitation. The vapor moving through the liquid keeps everything well mixed.
2. Equilibrium vapor is continuously generated for sampling. Long waits are not required. Equilibrium is minimally disturbed by sampling.
3. No preliminary degassing is required. This eliminates the major problem encountered in static methods.
4. No temperature controller is required. Temperature is controlled by the pressure. Since the bubble cap and pipe are surrounded by boiling liquid, the liquid and vapor samples are very nearly the same temperature. Temperature differences as measured by moving the thermocouple up and down the thermowell were less than 0.4 K.
5. Only one theoretical plate exists in the internal portion of the still. The internal still (pipe and bubble cap) is surrounded by liquid in the outer still which is very nearly uniform in temperature; therefore, the need to maintain the vapor space at a temperature slightly above that of the liquid is eliminated.
6. No vapor condensate is recirculated. This allows steady-state to be achieved as soon as boiling occurs.
7. Vapor samples are withdrawn as vapor. Those samples which would condense to two phases do not present problems.

Disadvantages of this still are:

1. The still cannot be used at low pressures. The pressure correction of the static head of liquid above the vapor-liquid interface is difficult to estimate accurately and is negligible at high pressures but significant at low pressures.
2. The dynamic nature of the still prevents its use with wide boiling mixtures. The more volatile component will pass the condenser, into the ballast tank. A second, colder condenser can be added to the top of the fixed condenser to help alleviate this problem.
3. Knowledge of the liquid composition depends on analysis. The unknown vapor volume prevents the use of a mass balance to convert the charge composition to the liquid composition.

PTx Measurements

The apparatus used for making PTx measurements as shown in Figure 2 was simple in construction and in operation.

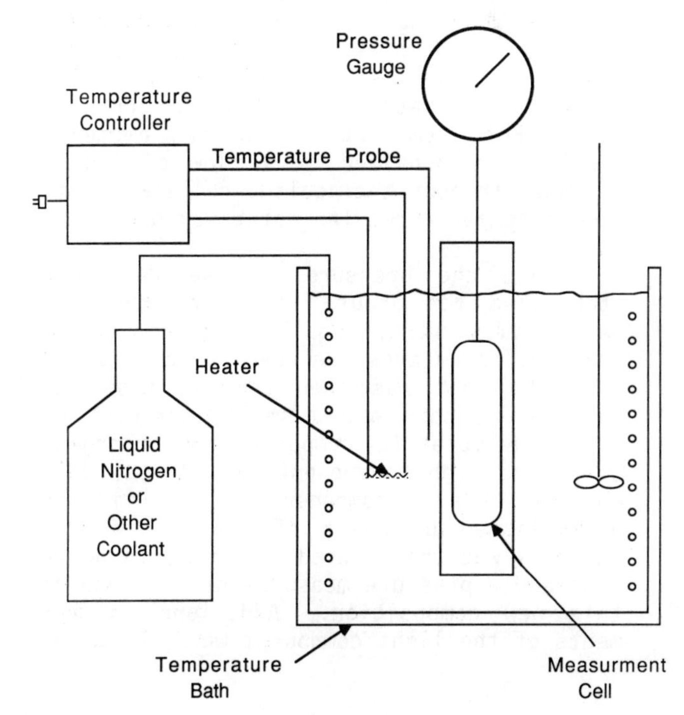

Figure 2. Schematic of static cell used in PTx measurements.

The measurement cell was a 300 ml stainless steel cylinder connected to a pressure gauge of known dead volume. The cell was mounted in such a way as to allow agitation. Pure samples of the two compounds of the binary mixture to be studied were first placed in 300 ml stainless steel charging cylinders. The samples were then degassed in a manner similar to that recommended by Ronc and Ratcliff (1). The filled cylinders were hooked up to a vacuum pump and immersed in liquid nitrogen. As the samples were cooled and frozen, the vapor above the samples in the cylinders was evacuated. The cylinder was warmed to room temperature and the process repeated 5 to 6 times.

The total pressure cell was evacuated and a few gm of the less volatile compound were charged into the cell. The amount of charge was determined by weighing the charging cylinder before and after charging. The transfer line connecting the charging cylinder and the total pressure cell was a 0.159 cm (1/16 in.) tubing approximately 2.5 cm in length. After closing the charging cylinder, the transfer line was heated to vaporize the liquid remaining in the transfer line. The small amount of vapor left in this line represented less than 0.01% of the amount charged and was considered negligible.

The charging cylinder was disconnected and the total pressure cell was lowered into a temperature bath maintained at a constant temperature by a feedback controller. For temperatures below ambient, cooling was provided by a flow of liquid nitrogen through the cooling coils while at higher temperatures tap water was used.

Once the pressure in the cell had stabilized for about fifteen minutes, it was assumed that equilibrium had been reached, and a pressure reading was recorded. The cell was then removed from the temperature bath and a small increment of the more volatile component was charged. Since the light component was always added to the heavy component, charging was accomplished by means of vapor pressure. The cell was then placed in the temperature bath and a pressure measurement was made at this new composition. Additional increments of the light component were added so as to give approximately equally spaced points across the composition range.

The composition range was studied in two halves. One half was measured by limiting the initial charge of heavy component to about 2 ml. This allowed the range of 40 mole % to 99 mole % of the light component to be measured. The 1 mole % to 60 mole % range was studied by charging 100 ml of the heavy component initially. The degree of agreement of the pressures at overlapping compositions was a test for the consistency of the data.

The total pressure method has several advantages:

1. No liquid or vapor sample analyses are required. This eliminates sampling and analytical errors.
2. The system is closed. The charge composition can be determined accurately by weight. A mass balance is used to determine liquid composition.
3. Basic measurements consist only of temperature, charge composition, and pressure.

The disadvantages are as follows:

1. Complete degassing is required. Degassing errors are usually hard to detect, yet create serious errors in the data reduction.
2. No easy check is available to determine if equilibrium has been reached.
3. Compositions of the vapor and liquid depend on the model used to fit the measured data.
4. The method can only be used for binary data.
5. An equation of state used must be able to predict vapor density accurately, otherwise the mass balance will fail to accurately convert charge composition to liquid composition.

Sample Analysis

Vapor and liquid samples collected using the reflux cell were analyzed using a Hewlett-Packard 5830A gas chromatograph with a flame ionization detector with an accompanying 18850A GC terminal. The chromatograph recorded and automatically integrated the peaks as they eluted from the column. Peak areas and area per cents were printed by the terminal at the end of every analysis.

A chromatographic response factor for each mixture for the flame ionization detector was determined as a function of composition. Standards of known composition were prepared gravimetrically. The standards were then analyzed in the same manner as the corresponding samples from the still. A response factor was calculated using Equation (1),

$$\dot{F} = (x_1/x_2)/(area_1/area_2) \quad (1)$$

where x_1 and x_2 are mole fractions of components 1 and 2 and $area_1$ and $area_2$ are the areas of the corresponding chromatographic peaks. Standards of several different compositions were analyzed so as to include the entire composition range.

The ratio of the chromatographic areas, $(area_1/area_2)$, determined for the samples were multiplied by the response factor, F, to convert the area ratio to a mole ratio. The mole ratio was combined with Equation (2),

$$x_1 + x_2 = 1 \quad (2)$$

to give the mole fractions of each sample analyzed using the chromatograph. Reproducibility within a sample was approximately 0.5% while reproducibility between samples was about 1%.

Helium was used as the carrier gas for all systems analyzed by the chromatograph. All columns were 0.3175 cm (1/8 in) stainless steel tubing packed with the appropriate chromatographic packing. The sample handling was different for each system and the technique was determined by trial and error.

The methanol + pyridine samples were drawn through a septum into a closed 2.5 ml vial then diluted 3:1 with n-propanol, except in cases where the total pressure in the still was below atmospheric, then the samples were drawn into a gas tight syringe then placed into vial and diluted. A 0.02 µl sample was then injected into a 2.4 m column packed with Porapak PS, 80-100 mesh packing. The flow rate of the helium was set at 40 ml/min and the oven was temperature programmed to rise from 423 K to 463 K through the course of the analysis.

The n-hexane + ortho-chlorotoluene samples were drawn in the same manner as the methanol + pyridine samples. Toluene was used as the diluent in a 3:1 ratio. The column was 0.30 m long and packed with Chromasorb 102, 80-100 mesh. The helium flow rate was 30 ml/min. The oven temperature was programmed from 423 K to 463 K.

The 1-butene + MTBE samples were two phase at room conditions so a special technique was used to collect and inject the samples. A 1.5 liter cylinder was evacuated and the sample was drawn directly into it. The samples were diluted 5:1 with nitrogen which also served to pressurize the container. The cylinder was heated to approximately 423 K by use of a heating tape which was wrapped around the cylinder. This insured that the sample existed as a single phase. The cylinder was connected via a 0.159 cm (1/16 in.) tube to a 0.03 ml heated gas sample valve. The pressure inside the cylinder pushed the gas through the valve for sampling. The helium flow rate was 30 ml/min and the column was 2.4 m in length and contained Porapak PS, 80-100 mesh. The oven temperature was constant at 423 K.

Propylene + 1,3-butadiene samples were drawn directly into a 50 ml gas tight syringe. The needle was placed in a rubber stopper to prevent any leakage. The samples were injected, without dilution, into the 0.03 ml gas sample valve. The column was 2.4 m long and contained Porapak PS, 80-100 packing. The flow rate was 40 ml/min and the temperature was constant at 363 K.

Vinyl acetate + n-hexane samples were drawn through a septum into a closed 2.5 ml vial then diluted 2:1 with acetone. A 0.02 µl sample was injected into the chromatograph which was a column 0.91 m long and packed with 80-100 mesh Porapak

QS. The flow rate was 30 ml/min and the temperature constant at 443 K.

Vinyl acetate + methyl ethyl ketone samples were withdrawn from the still in the same manner as the vinyl acetate + hexane samples and diluted 3:1 with toluene. The column was packed with Porapak QS, 80-100 mesh and was 0.91 m long. The helium flow rate was 30 ml/min and the temperature was programmed from 443 K to 473 K.

The phenol + methyl ethyl ketone samples were withdrawn using a gas tight syringe and diluted 3:1 with toluene. The column was 0.30 m long and was packed with 80-100 mesh Chromasorb 102. The carrier gas flow rate was 50 ml/min and the temperature programmed to go from 403 K to 503 K.

The acetaldehyde + 1-pentene samples were withdrawn through a septum into a 2.5 ml vial and diluted 3:1 with ethanol. The helium flow rate was 30 ml/min and the temperature programmed from 423 K to 448 K.

No sample analysis was performed on the remaining two systems as they were measured using pTx and calorimetric measurements.

Liquid-Liquid Measurements

The liquid-liquid compositions of the carbon disulfide + methanol system were determined in two ways. At 233 K and 253 K, the total pressure apparatus was used and the constant pressure region was assumed to be the two phase region and the abscissa of the break points in the pressure curve were assumed to be the compositions of the two phases as shown in Figures 3 and 4. The measurements at 273 K and 293 K were made in a flow calorimeter by colleagues at B.Y.U. (2).

Measurements of the Vapor Pressure of Pure Components

The vapor pressures of methanol, ortho-chlorotoluene, 1-butene, methyl

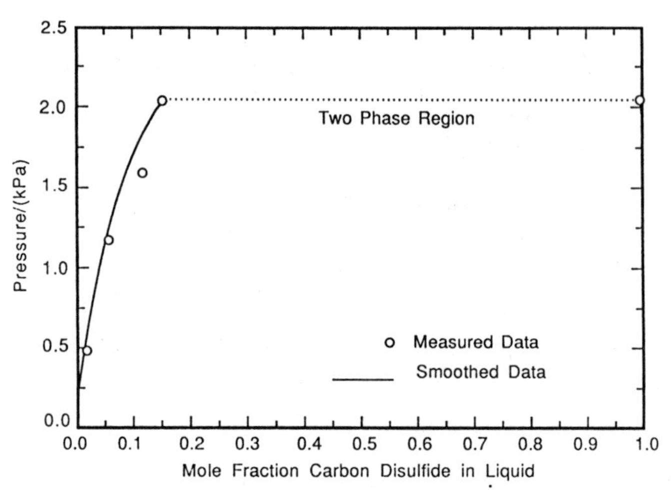

Figure 3. Total pressure measurements with smoothed fit using Wilson equation for the carbon disulfide + methanol system at 333.2 K.

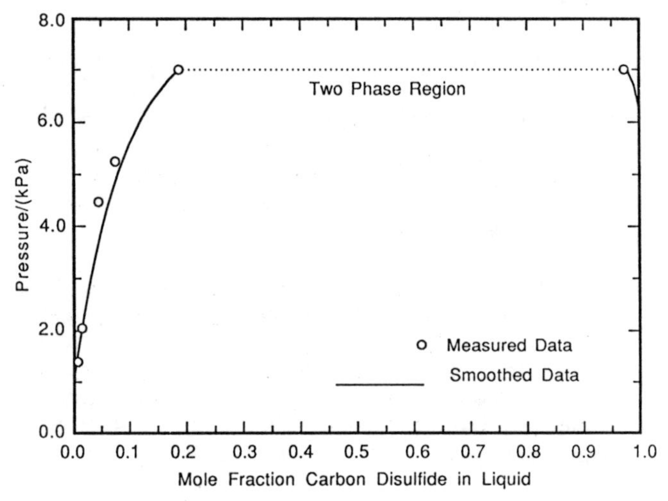

Figure 4. Total pressure measurements with smoothed fit using Wilson equation for the carbon disulfide + methanol system at 353.2 K.

tertiary-butyl ether, propylene, 1,3-butadiene, pyridine, vinyl acetate, n-hexane, methyl ethyl ketone, acetaldehyde, and 1-pentene were measured as a function of temperature using the bubble cap still. The pure component was added to the cell and the pressure was set; the heaters were turned on and the fluid in the cell was allowed to reach the equilibrium temperature. The temperature and pressure were recorded and the pressure increased and the system allowed to come to equilibrium again.

This method of measuring the vapor pressure eliminates the need for degassing as the system is boiling and all the light components escape out the reflux condenser. The relative errors in pressures measured by this method are greatest at low pressures since the correction for the static head above the bubble cap becomes more significant when compared to the total pressure. If significant heavy impurities are in the liquid, then they will remain in the still and influence the vapor phase. Both these errors would cause the measured vapor pressures to be lower than the real pure component vapor pressures.

Chemicals Used

The methanol was absolute methanol Mallincrodt analytical reagent, 99.8% pure and was used without further purification. The carbon disulfide was Mallincrodt analytical reagent, 99.5% pure and used without further purification. The pyridine was MCB reagent grade, 99+% pure and also used without purification. The carbonyl sulfide was Matheson gas of 97.5% purity. The carbonyl sulfide was distilled to remove the heavy components and degassed to remove the more volatile components. The vinyl acetate was Fisher reagent grade inhibited 99+% pure and was used without further purification. The n-hexane was Phillips 99+% and was used without further purification. The methyl ethyl ketone was Fisher reagent grade 99+% and used as delivered. The phenol was Aldrich GOLD LABEL 99+% and was used without further purification. It was kept under a dry nitrogen atmosphere. The propylene was Matheson CP grade 99.2% pure and was used as delivered. The 1,3-butadiene was Matheson CP grade 99.4% pure and containing 115 p. p. m. tertiary-butylcatechol as an inhibitor and used without further purification. The 1-butene used was Matheson CP grade 99.2% pure and used without further purification. The methyl tertiary-butyl ether (MTBE) was obtained from Exxon. It was guaranteed 98% pure by weight. It was redistilled before use to give a product 99+% pure. The 1-pentene was Aldrich 99+% pure and used without further purification. The acetaldehyde was Fisher 99+% and used as delivered. The ortho-chlorotoluene was Aldrich 98% pure and was distilled before use so the estimated purity was 99+%.

EXPERIMENTAL RESULTS AND DISCUSSION

Vapor Pressures of Pure Compounds

The vapor pressures of n-hexane, ortho-chlorotoluene, methanol, pyridine, 1-butene, MTBE, propylene, 1,3-butadiene, acetaldehyde, methyl ethyl ketone, vinyl acetate, and 1-pentene as measured in this project are listed in Tables 1 to 12. As can be seen, the agreement between the measured pressures and the pressures reported in the literature is generally good with the exception of the vapor pressure of acetaldehyde where the measured vapor pressures were consistently lower than the vapor pressures reported in the literature. It was found that using the measured vapor pressures of acetaldehyde in the reduction of the vapor liquid equilibrium data for the 1-pentene + acetaldehyde system gave more thermodynamically consistent results than using the literature values. The vapor pressures of the methyl tertiary - butyl ether and ortho - chlorotoluene were fitted with the Antoine equation and the parameters are given in Tables 9 and 12.

PTx Data

The PTx data were reduced to composition data using Equation 3,

$$\gamma_i x_i P_i^o (f_i^o/P_i^o)\ \phi_i = y_i P\ (f_i/P_i) \qquad (3)$$

Table 1. Measured vapor pressure for methanol.

Temperature, K	Measured Pressure, kPa	Literature Value(3,4), kPa
331.9	78.24	81.11
335.4	89.95	93.57
340.1	108.91	112.72
345.5	137.17	138.56
366.9	288.13	291.96
376.6	387.7	395.8
378.7	417.4	421.7
382.0	460.1	465.1
382.3	466.0	469.3
386.0	518.4	522.4
386.4	527.3	528.4
391.6	610.0	611.8
395.9	689.3	688.2
400.3	773.7	773.9
403.9	856.5	850.1
406.0	899.5	897.2
409.2	982.3	972.8
411.5	1044.3	1030.1
413.2	1082.2	1074.2
413.6	1096.0	1084.8

Table 2. Measured vapor pressure for pyridine.

Temperature, K	Measured Pressure, kPa	Literature Value(3,4), kPa
336.1	16.34	16.75
341.4	20.47	20.47
351.2	29.71	29.98
359.1	39.29	39.98
362.5	44.53	44.32
372.6	62.3	61.8
378.1	74.2	74.6
381.0	86.2	86.6
384.1	89.2	89.5
387.8	100.5	100.6
389.4	112.0	--
392.7	122.4	121.9
397.4	136.1	--
401.4	153.4	--
405.9	174.0	--
410.4	196.5	--
414.8	222.3	--
419.9	253.3	--
426.1	289.5	--
428.7	311.9	--
432.7	342.9	--
436.6	373.9	--
441.1	411.9	--
443.3	434.3	--

Table 3. Measured vapor pressure for vinyl acetate.

Temperature, K	Measured Pressure, kPa	Literature Value(3,4), kPa
328.1	53.42	54.09
334.1	67.21	67.26
340.9	84.44	85.15
346.0	99.95	100.90
353.0	124.07	126.15
359.0	151.7	151.5
366.4	189.6	188.1
371.9	220.6	219.5
379.7	272.3	270.6
387.2	330.9	328.0
392.7	379.1	375.6
398.7	432.5	433.3

Table 4. Measured vapor pressure for n-hexane.

Temperature, K	Measured Pressure, kPa	Literature Value(3,4), kPa
313.2	34.81	37.25
329.8	68.93	68.07
337.1	87.89	86.80
342.9	101.67	104.40
345.7	112.01	113.83
352.7	137.9	140.3
353.2	141.0	142.4
358.5	165.4	165.7
362.4	186.1	184.7
368.4	218.9	217.1
373.5	248.2	247.8
380.9	298.1	298.3
387.3	349.8	347.8
393.2	398.4	398.7
393.7	405.0	403.2
398.7	451.5	450.9
433.2	903.0	900.9

Table 5. Measured vapor pressure for methyl ethyl ketone.

Temperature, K	Measured Pressure, kPa	Literature Value(3,4), kPa
332.1	48.25	50.32
333.1	51.70	52.17
335.8	55.14	57.42
340.0	65.48	66.44
348.0	86.16	86.76
352.6	99.95	100.54
361.8	130.97	133.32
363.9	141.31	141.87
372.7	179.22	182.57
374.8	194.73	193.52
379.2	220.6	218.1
383.4	244.7	243.8
386.0	263.7	260.9
391.0	299.9	296.3
393.6	317.1	316.1
395.8	337.8	333.7
397.4	351.5	347.0

Table 6. Measured vapor pressure for propylene.

Temperature, K	Measured Pressure, kPa	Literature Value(3,4), kPa
268.9	510.1	513.9
270.1	523.9	533.2
271.9	561.6	563.1
273.8	589.4	596.1
275.7	623.8	630.4
279.2	696.2	697.4
286.0	847.8	842.4
297.8	1140.8	1144.8
298.7	1151.1	1170.7
303.5	1309.7	1316.2
310.1	1506	1537
317.2	1782	1804
317.6	1785	1820
321.2	1940	1968
323.7	2054	2077
326.7	2185	2212
329.5	2320	2344
330.5	2385	2392
334.1	2550	2574

Table 7. Measured vapor pressure for 1,3-butadiene.

Temperature, K	Measured Pressure, kPa	Literature Value(3,4), kPa
267.2	87.89	95.16
267.7	89.61	97.82
268.7	93.06	100.93
271.5	103.40	112.44
273.7	115.46	122.18
281.6	151.7	162.6
287.2	186.1	197.0
294.2	243.0	247.6
298.5	279.2	283.1
306.2	353.3	356.3
314.4	444.6	448.8
320.7	517.0	531.0
323.4	558.3	569.5
326.6	610.0	617.6
329.2	651.4	658.9
331.4	691.0	695.3
333.4	730.7	729.7

Table 8. Measured vapor pressure for 1-butene.

Temperature, K	Measured Pressure, kPa	Literature Value(3,4), kPa
288.5	215.4	218.9
289.7	227.5	227.6
292.2	244.7	246.6
294.2	258.5	262.7
297.6	287.8	291.8
299.8	306.7	311.9
306.0	370.5	374.0
314.3	465.3	471.2
317.2	499.7	509.1
322.0	570.4	576.8
325.3	620.4	627.0
329.2	685.9	690.2
334.2	768.6	778.0
336.5	809.9	520.9
340.4	889.2	897.6
343.5	963.3	962.0
345.6	1006.4	1007.5
348.4	1068.4	1070.5
349.8	1106.3	1103.0
352.0	1158.0	1155.5
353.3	1192.5	1187.3

Table 9. Measured vapor pressure for methyl tertiary-butyl ether.

Temperature, K	Measured Pressure, kPa	Correllated Pressure*, kPa
292.3	24.47	24.58
293.6	26.19	26.13
294.1	27.26	26.88
295.0	28.26	27.88
296.2	29.64	29.45
299.0	34.47	33.38
301.5	37.22	37.22
304.3	42.05	41.91
306.4	45.84	45.72
315.4	64.11	65.07
331.1	114.4	112.7
335.0	129.2	127.7
339.0	142.34	144.55
348.5	191.6	190.9
353.4	217.5	218.5
355.8	232.3	233.0
359.2	255.0	254.7

*Fitted to Antoine equation: $\ln P = A - B/(T + C)$,
where P is pressure in kPa, A = 12.1667, B = 1699.76, C = -102.645

Table 10. Measured vapor pressure for 1-pentene.

Temperature, K	Measured Pressure, kPa	Literature Value(3,4), kPa
304.9	105.12	107.49
309.0	120.63	123.43
313.6	141.31	143.45
318.6	165.43	167.48
322.7	189.56	190.36
327.6	218.9	220.1
333.0	249.9	256.7
336.7	282.6	284.4
342.6	334.3	333.2
348.3	386.0	386.1
353.0	442.9	434.2
357.5	494.6	484.4
362.5	561.8	545.1
366.9	620.4	603.1
371.7	694.5	671.5
377.8	798.2	766.5
383.5	863.4	864.0

Table 11. Measured vapor pressure for acetaldehyde.

Temperature, K	Measured Pressure, kPa	Literature Value(3,4), kPa
289.2	80.99	85.67
293.1	93.06	99.45
298.1	105.12	119.61
301.3	122.35	134.12
306.1	144.75	158.43
311.0	170.6	186.7
315.8	198.2	218.0
319.6	224.0	245.6
322.4	246.4	267.5
327.6	286.1	312.3
333.3	344.7	367.8
337.5	387.7	413.2
343.0	449.8	478.9
347.6	508.4	539.7
353.0	582.5	618.2
358.1	663.5	699.9
363.3	746.2	791.1
368.3	825.4	886.7
373.5	946.1	994.9
378.0	1049.5	1096.0
382.9	1161.5	1214.4

Table 12. Measured vapor pressure for ortho-chlorotoluene.

Temperature, K	Measured Pressure, kPa	Correllated Pressure*, kPa
382.4	23.71	24.36
388.4	29.23	29.43
398.2	39.70	39.59
403.6	46.80	46.34
412.1	59.97	58.89
421.5	77.89	75.92
426.5	87.89	85.63
428.9	94.78	92.02
433.2	106.15	102.60
436.6	110.29	111.65
442.5	129.2	128.9
450.3	156.8	155.0
455.2	174.1	173.5
460.8	199.9	196.8
468.2	232.6	231.3
474.9	268.8	266.7
477.6	282.6	282.1

*Fitted to Antoine equation: lnP = A - B/(T + C),
where P is pressure in kPa, A = 15.7455, B = 4929.06, C = 10.324

where γ_i is the activity coefficient; x_i is the liquid mole fraction; y_i is the vapor mole fraction; P_i^o is the pure component vapor pressure; P is the total pressure; f_i^o/P_i is the pure component fugacity coefficient; f_i/P_i is the vapor phase fugacity coefficient; and ϕ_i is the Polynting correction. The subscript i refers to component i in the mixture. The activity coefficient was calculated using the modified Wilson Equation given as Equation 4,

$$\ln\gamma_i = c\left[-\ln(x_1 + \Lambda_{12}x_2) - \frac{x_1}{x_1 + \Lambda_{12}x_2} - \frac{\Lambda_{21}x_2}{x_2 + \Lambda_{21}x_1} + 1\right]$$

$$\ln\gamma_2 = c\left[-\ln(x_2 + \Lambda_{21}x_1) - \frac{x_2}{x_2 + \Lambda_{21}x_1} - \frac{\Lambda_{12}x_1}{x_1 + \Lambda_{12}x_2} + 1\right] \quad (4)$$

where Λ_{12}, Λ_{21}, and c are adjustable parameters. Several investigators (1,5,6) have shown that the parameter, c, which modifies the original two parameter Wilson Equation, is effective in predicting heterogeneous mixtures.

The fugacity coefficients were calculated using a modified Redlich-Kwong equation of state which is defined by Equations (5) to (12) where f is the fugacity, P the total pressure, V the molar volume, T_c the critical temperature, P_c the critical pressure, and ω the acentric factor.

$$z = \frac{1}{1 - B_T/V} - \frac{F_T B_T/V}{1 + B_T/V} \tag{5}$$

where

$$z = \frac{PV}{RT} \tag{6}$$

$$B_i = 0.0865 \frac{RT_{c_i}}{P_{c_i}} \tag{7}$$

$$F_i = 4.934\, T_{r_i}^{-0.22} \{1 + (1.2303 + 1.91\,\omega_i - 0.52\,\omega_i^2)(\frac{1}{T_{r_i}} - 1) + 0.0006 \exp[-2(\frac{1}{T_{r_i}} - 1)^2]\} \tag{8}$$

$$B_T = \Sigma y_i B_i \tag{9}$$

$$F_T = \Sigma y_i F_i \tag{10}$$

$$T_{r_i} = \frac{T}{T_{c_i}} \tag{11}$$

Once z and V have been obtained by solution of the above equations, the fugacity coefficient coefficient can be obtained by using Equation (12).

$$\ln(\frac{f_i}{P_i}) = \ln(1 - \frac{B_T}{V}) - F_i \ln(1 + \frac{B_T}{V}) + \frac{B_i\,(z - 1)}{B_T} - \ln z \tag{12}$$

The values of Λ_{12}, Λ_{21}, and c in Equation (3) are were varied until a minimum error among the measured and predicted total pressures was obtained for a given isotherm. Getting the total pressure from a given set of Wilson equation parameters was in itself an iterative process requiring convergence of the material balance.

Since the modified Wilson Equation gives thermodynamically consistent activity coefficients, the agreement between the fitted pressures and measured pressures gives an estimate of the consistency, provided that all assumptions employed are verified. The data for the PTx measurements together with the vapor compositions and the pressures predicted by the correlating equations are given in Tables 13 to 19. All of the systems show a positive deviation from ideality. The carbon disulfide + methanol system exhibited liquid-liquid equilibrium so the range of useful pressure measurements was limited. The scatter in the measurements indicates that the data for n-hexane + ortho-chlorotoluene system at 353 K are the least accurate.

Table 13. PTx data for the carbonyl sulfide + methanol system at 233.2 K.

Mole Fraction Carbonyl Sulfide			Pressure, kPa		Activity Coefficients	
Charge	Liquid	Calculated Vapor	Measured	Calcuated	COS	MeOH
0.0751	0.0670	0.9960	44.46	45.42	4.361	1.007
0.1648	0.1523	0.9979	83.41	83.84	3.516	1.034
0.2307	0.2174	0.9984	103.40	103.71	3.033	1.067
0.2862	0.2735	0.9986	116.49	116.37	2.697	1.110
0.3294	0.3175	0.9987	124.07	124.14	2.474	1.151
0.3710	0.2011	0.9983	99.60	99.33	3.143	1.059
0.4157	0.4061	0.9989	137.17	135.59	2.107	1.261
0.4921	0.4848	0.9990	144.41	142.41	1.851	1.400
0.5056	0.3635	0.9988	133.72	130.68	2.271	1.203
0.6123	0.6084	0.9991	150.61	149.07	1.541	1.748
0.6191	0.5309	0.9990	149.58	145.37	1.724	1.506
0.7184	0.6724	0.9991	156.47	151.22	1.414	2.038
0.7912	0.7673	0.9991	158.20	153.36	1.256	2.772
0.9001	0.8951	0.9992	159.23	155.17	1.089	5.803
0.9223	0.9194	0.9992	161.30	155.53	1.063	7.388

Table 14. PTx data for the carbonyl sulfide + methanol system at 253.2 K.

Mole Fraction Carbonyl Sulfide			Pressure, kPa		Activity Coefficients	
Charge	Liquid	Calculated Vapor	Measured	Calcuated	COS	MeOH
0.0287	0.0236	0.9741	38.26	38.33	4.922	1.001
0.0733	0.0622	0.9892	89.95	90.23	4.418	1.006
0.1152	0.1001	0.9927	133.04	132.06	3.998	1.015
0.1708	0.1527	0.9947	181.29	178.27	3.512	1.034
0.2265	0.2076	0.9957	217.48	215.22	3.099	1.063
0.3262	0.3092	0.9967	266.76	262.35	2.514	1.143
0.4124	0.3989	0.9971	295.02	288.70	2.134	1.251
0.5167	0.5079	0.9974	316.73	308.83	1.786	1.450
0.5644	0.3293	0.9968	268.83	269.26	2.419	1.164
0.7797	0.7247	0.9977	303.98	327.69	1.323	2.380
0.8356	0.8073	0.9978	331.21	330.95	1.199	3.298
0.8762	0.8610	0.9979	337.07	332.55	1.129	4.468
0.9137	0.9068	0.9979	337.41	333.88	1.076	6.471
0.9381	0.9347	0.9980	340.86	334.89	1.047	8.924
0.9540	0.9523	0.9981	342.24	335.76	1.030	11.708

Table 15. PTx data for the carbonyl sulfide + methanol system at 273.2 K.

Mole Fraction Carbonyl Sulfide			Pressure, kPa		Activity Coefficients	
Charge	Liquid	Calculated Vapor	Measured	Calcuated	COS	MeOH
0.1134	0.0857	0.9818	218.85	218.53	4.150	1.011
0.1789	0.1430	0.9877	320.53	319.43	3.595	1.030
0.2345	0.1957	0.9901	398.073	390.76	3.181	1.056
0.3662	0.3320	0.9928	523.87	510.85	2.407	1.167
0.4947	0.4724	0.9939	591.08	575.99	1.888	1.375
0.5555	0.2743	0.9920	472.17	469.09	2.693	1.111
0.5866	0.5727	0.9944	620.37	602.25	1.621	1.625
0.6664	0.2825	0.9921	537.66	475.73	2.649	1.118
0.7514	0.6597	0.9946	635.88	616.68	1.438	1.972
0.8077	0.7583	0.9949	653.11	626.95	1.270	2.677
0.9097	0.9012	0.9953	661.73	635.84	1.082	6.133
0.9327	0.9284	0.9954	663.45	637.45	1.053	8.221

Table 16. PTx data for the carbonyl sulfide + methanol system at 293.2 K.

Mole Fraction Carbonyl Sulfide			Pressure, kPa		Activity Coefficients	
Charge	Liquid	Calculated Vapor	Measured	Calcuated	COS	MeOH
0.0483	0.0315	0.9188	158.54	161.05	4.812	1.001
0.0706	0.0471	0.9422	220.58	223.89	4.367	1.003
0.0972	0.0665	0.9565	299.85	302.53	4.367	1.007
0.1348	0.0955	0.9673	404.97	403.89	4.046	1.013
0.1815	0.1342	0.9746	523.87	520.06	3.672	1.026
0.2571	0.2036	0.9808	703.09	683.89	3.126	1.060
0.3626	0.3129	0.9849	889.20	857.80	2.496	1.147
0.4643	0.4271	0.9870	1008.11	967.90	2.033	1.294
0.5548	0.5306	0.9981	1068.4	1028.5	1.725	1.505
0.9369	0.9120	0.9904	1109.8	1108.2	1.0705	6.820
0.9701	0.9666	0.9917	1125.3	1116.9	1.018	15.603

Table 17. PTx data for the n-hexane + ortho-chlorotoluene system at 353.2 K.

Mole Fraction n-Hexane			Pressure, kPa		Activity Coefficients	
Charge	Liquid	Calculated Vapor	Measured	Calcuated	n-Hexane	OCT
0.2350	0.2323	0.8623	52.39	52.45	1.417	1.031
0.2838	0.2811	0.8843	58.59	59.55	1.359	1.046
0.3472	0.3448	0.9051	73.41	68.09	1.293	1.070
0.3897	0.3875	0.9160	79.61	73.44	1.198	1.126
0.4560	0.4543	0.9298	94.09	81.34	1.198	1.126
0.4978	0.4964	0.9372	98.92	86.09	1.167	1.153
0.5397	0.5387	0.9438	107.19	90.73	1.139	1.183
0.5597	0.5560	0.9438	106.84	92.60	1.129	1.196
0.5958	0.5952	0.9518	114.77	96.79	1.107	1.229
0.6230	0.6225	0.9554	115.80	99.68	1.093	1.253
0.6516	0.6513	0.9591	120.63	102.71	1.079	1.281
0.7910	0.7490	0.9707	109.25	113.05	1.041	1.394
0.8329	0.7490	0.9707	116.49	113.05	1.041	1.394
0.8630	0.8452	0.9817	120.63	123.60	1.016	1.535
0.9015	0.8963	0.9875	129.93	129.49	1.007	1.625
0.9333	0.9294	0.9914	133.04	133.46	1.003	1.691
0.9482	0.9459	0.9934	135.86	135.48	1.002	1.725
0.9643	0.9638	0.9955	137.86	137.72	1.001	1.765
0.9663	0.9654	0.9957	139.93	137.93	1.001	1.769
0.9690	0.9683	0.9961	140.27	138.29	1.001	1.775
0.9756	0.9754	0.9970	140.96	139.20	1.000	1.792
0.9808	0.9806	0.9976	139.93	139.87	1.000	1.804
0.9831	0.9829	0.9979	140.96	140.17	1.000	1.809
0.9849	0.9829	0.9981	142.00	140.41	1.000	1.814
0.9866	0.9865	0.9983	142.69	140.63	1.000	1.818
0.9880	0.9879	0.9985	144.06	140.81	1.000	1.821

Table 18. PTx data for the carbon disulfide + methanol system at 233.2 K.

Mole Fraction Carbon Disulfide			Pressure, kPa		Activity Coefficients	
Charge	Liquid	Calculated Vapor	Measured	Calcuated	CS_2	MeOH
0.0186	0.0185	0.6917	0.48	0.60	12.005	1.001
0.0574	0.0572	0.8548	1.17	1.24	9.862	1.009
0.1173	0.1170	0.9052	1.58	1.83	7.516	1.035

Table 19. PTx data for the carbon disulfide + methanol system at 253.2 K.

Mole Fraction Carbon Disulfide			Pressure, kPa		Activity Coefficients	
Charge	Liquid	Calculated Vapor	Measured	Calcuated	CS_2	MeOH
0.0078	0.0077	0.3649	1.38	1.57	11.993	1.000
0.0157	0.0156	0.5286	2.07	2.10	11.468	1.001
0.0441	0.0439	0.7351	4.48	3.64	9.843	1.005
0.0752	0.0748	0.8057	5.24	4.85	8.434	1.015

PTxy Data

The vapor-liquid equilibrium data for the eight systems were obtained using the bubble cap reflux still. Direct measurements were made of the liquid compositions, vapor compositions, and the pressures. This allowed the direct calculation of activity coefficients. Activity coefficients are very sensitive to errors in all the measurements. Van Ness (7) has shown that the commonly applied area test for activity coefficients is of little use. He recommends testing consistency of PTxy data by using three properties to predict the fourth.

As a check on consistency, a modified Wilson Equation was fitted to all PTxy data. The technique required use of the same equations as used in the reduction of the PTx data, Equations (3) through (12). Two major differences existed between the data reduction methods used to fit the PTx data and the PTxy data. The first difference was that no mass balance was required for the reduction of the PTxy data as the x values were measured. The second was that the parameters were adjusted to fit the measured vapor mole fractions rather than the total pressure.

The data together with the fitted vapor compositions, pressures, and smoothed activity coefficients are reported in Tables 20 to 27. The generally good

Table 20. PTxy data for the methanol + pyridine system.

Temp., K	Mole Fraction Methanol			Pressure, kPa		Activity Coefficients Calculated	
	Measured Liquid	Measured Vapor	Calculated Vapor	Measured	Calculated	MeOH	Pyridine
353.2	0.156	0.514	0.526	53.4	58.2	1.091	1.003
353.2	0.349	0.750	0.754	81.0	88.2	1.054	1.015
353.2	0.495	0.841	0.843	102.7	109.9	1.033	1.030
353.2	0.832	0.961	0.961	153.4	158.7	1.004	1.090
353.2	0.982	0.996	0.996	175.8	180.8	1.000	1.130
373.2	0.150	0.496	0.500	103.4	110.1	1.068	1.002
373.2	0.357	0.750	0.751	163.7	171.4	1.039	1.012
373.2	0.589	0.876	0.881	226.8	238.0	1.016	1.032
373.2	0.830	0.958	0.960	289.5	306.9	1.003	1.065
373.2	0.982	0.995	0.996	334.3	351.1	1.000	1.093
393.2	0.0295	0.138	0.146	132.3	132.2	1.089	1.000
393.2	0.261	0.642	0.652	251.6	256.7	1.051	1.006
393.2	0.592	0.872	0.877	424.6	426.7	1.016	1.032
393.2	0.830	0.954	0.958	544.5	549.3	1.003	1.065
393.2	0.982	0.995	0.996	622.1	629.7	1.000	1.093
413.2	0.0290	0.133	0.138	225.7	223.2	1.089	1.000
413.2	0.348	0.725	0.724	499.7	503.1	1.040	1.011
413.2	0.589	0.864	0.869	711.4	710.0	1.016	1.032
413.2	0.982	0.995	0.996	1061.5	1057.5	1.000	1.092

Table 21. PTxy data for the n-hexane + vinyl acetate system.

Temp., K	Mole Fraction n-Hexane			Pressure, kPa		Activity Coefficients Calculated	
	Measured Liquid	Measured Vapor	Calculated Vapor	Measured	Calculated	n-Hexane	Vinyl Acetate
333.2	0.0155	0.0154	0.0481	67.2	67.4	2.749	1.000
333.2	0.149	0.266	0.297	82.7	81.2	2.118	1.024
333.2	0.389	0.483	0.486	93.1	91.2	1.483	1.167
333.2	0.706	0.651	0.650	91.3	92.3	1.104	1.672
333.2	0.914	0.839	0.836	82.7	84.6	1.009	2.466
353.2	0.0151	0.0391	0.0398	129.2	130.4	2.421	1.000
353.2	0.147	0.241	0.268	153.4	152.1	1.939	1.020
353.2	0.403	0.466	0.476	170.6	169.7	1.393	1.156
353.2	0.706	0.656	0.651	167.2	170.3	1.090	1.572
353.2	0.913	0.844	0.844	155.1	156.3	1.008	2.198
373.2	0.0133	0.0334	0.0317	239.5	232.1	2.255	1.000
373.2	0.142	0.241	0.245	274.0	265.7	1.939	1.012
373.2	0.404	0.458	0.465	298.1	295.0	1.355	1.143
373.2	0.712	0.653	0.653	155.1	156.3	1.008	2.198
373.2	0.902	0.832	0.832	270.6	271.6	1.009	1.000
393.2	0.0139	0.0293	0.0278	396.3	386.2	1.992	1.000
393.2	0.133	0.209	0.209	444.6	426.8	1.710	1.012
393.2	0.389	0.428	0.435	487.7	466.8	1.315	1.111
393.2	0.712	0.651	0.652	477.3	461.7	1.067	1.438
393.2	0.905	0.845	0.841	442.9	423.9	1.008	1.835

Table 22. PTxy data for the vinyl acetate + methyl ethyl ketone system.

Temp., K	Mole Fraction Vinyl Acetate			Pressure, kPa		Activity Coefficients Calculated	
	Measured Liquid	Measured Vapor	Calculated Vapor	Measured	Calculated	Vinyl Acetate	MEK
333.2	0.0288	0.0378	0.0378	53.4	52.8	1.069	1.000
333.2	0.187	0.230	0.229	58.6	55.4	1.048	1.002
333.2	0.545	0.598	0.596	60.3	60.4	1.015	1.021
333.2	0.766	0.796	0.796	60.3	62.9	1.004	1.042
333.2	0.986	0.989	0.988	62.0	65.0	1.000	1.072
353.2	0.0300	0.0381	0.0390	103.4	103.5	1.068	1.000
353.2	0.190	0.225	0.232	106.8	108.5	1.047	1.003
353.2	0.539	0.584	0.588	113.7	117.7	1.015	1.021
353.2	0.775	0.803	0.803	120.6	122.8	1.004	1.043
353.2	0.987	0.988	0.989	125.8	126.7	1.000	1.072
373.2	0.0302	0.0385	0.0381	187.8	186.8	1.048	1.000
373.2	0.177	0.213	0.212	196.5	194.3	1.035	1.002
373.2	0.544	0.587	0.590	212.0	210.8	1.011	1.015
373.2	0.773	0.802	0.801	220.6	219.6	1.003	1.031
373.2	0.986	0.988	0.988	229.2	227.0	1.000	1.050
393.2	0.0305	0.0381	0.0379	320.5	315.7	1.048	1.000
393.2	0.169	0.202	0.201	330.9	327.1	1.035	1.001
393.2	0.547	0.594	0.589	358.4	354.3	1.010	1.015
393.2	0.767	0.793	0.973	372.2	367.7	1.003	1.030
393.2	0.986	0.988	0.988	384.3	379.5	1.000	1.050

Table 23. PTxy data for the methyl ethyl ketone + phenol system.

Temp., K	Mole Fraction MEK			Pressure, kPa		Activity Coefficients Calculated	
	Measured Liquid	Measured Vapor	Calculated Vapor	Measured	Calculated	MEK	Phenol
393.2	0.0966	0.378	0.384	19.0	18.7	0.256	0.981
393.2	0.199	0.648	0.679	25.8	30.2	0.354	0.928
393.2	0.392	0.899	0.913	67.6	70.6	0.559	0.768
393.2	0.731	0.993	0.992	212.0	199.1	0.889	0.424
413.2	0.0953	0.342	0.345	37.9	37.5	0.301	0.984
413.2	0.376	0.830	0.877	107.5	115.7	0.588	0.804
413.2	0.732	0.989	0.988	334.3	322.5	0.907	0.472
433.2	0.0918	0.306	0.299	72.4	68.6	0.337	0.987
433.2	0.191	0.573	0.571	102.0	96.7	0.433	0.947
433.2	0.355	0.809	0.828	182.0	174.1	0.599	0.839
433.2	0.719	0.975	0.981	511.5	479.7	0.909	0.520
453.2	0.0901	0.29.6	0.298	127.5	124.7	0.440	0.991
453.2	0.190	0.548	0.552	170.9	171.0	0.528	0.963
453.2	0.344	0.780	0.788	279.9	277.2	0.662	0.887
453.2	0.718	0.963	0.972	739.3	708.2	0.926	0.607

Table 24. PTxy data for the propylene + 1,3-butadiene system.

Temp., K	Mole Fraction Propylene			Pressure, kPa		Activity Coefficients Calculated	
	Measured Liquid	Measured Vapor	Calculated Vapor	Measured	Calculated	Propylene	1,3-BD
273.2	0.0763	0.291	0.295	160.3	149.2	1.081	1.000
273.2	0.0880	0.323	0.328	160.3	154.8	1.079	1.001
273.2	0.282	0.652	0.655	251.6	245.8	1.048	1.007
273.2	0.565	0.854	0.855	370.5	376.5	1.017	1.030
273.2	0.686	0.904	0.906	444.6	433.0	1.009	1.044
273.2	0.971	0.993	0.993	568.7	570.9	1.000	1.089
293.2	0.0430	0.160	0.155	279.2	268.2	1.087	1.000
293.2	0.734	0.245	0.243	289.5	291.7	1.082	1.001
293.2	0.281	0.598	0.600	448.0	450.5	1.048	1.007
293.2	0.566	0.829	0.823	668.6	667.5	1.017	1.030
293.2	0.686	0.881	0.883	765.1	760.8	1.009	1.044
293.2	0.922	0.975	0.975	965.0	952.3	1.001	1.080
293.2	0.972	0.991	0.911	985.7	994.6	1.000	1.090
313.2	0.112	0.303	0.297	627.3	562.4	1.075	1.001
313.2	0.278	0.557	0.551	768.6	755.0	1.049	1.007
313.2	0.459	0.726	0.720	975.4	966.1	1.027	1.019
313.2	0.576	0.800	0.800	1094.3	1105.1	1.016	1.031
313.2	0.686	0.860	0.861	1227.0	1239.0	1.009	1.044
313.2	0.921	0.968	0.968	1571.6	1540.6	1.001	1.080
333.2	0.107	0.264	0.257	999.5	899.3	1.075	1.001
333.2	0.267	0.507	0.500	1197.7	1133.9	1.050	1.007
333.2	0.568	0.763	0.766	1661.2	1684.6	1.017	1.030
333.2	0.679	0.834	0.836	1895.6	1885.7	1.009	1.043
333.2	0.933	0.968	0.968	2412.6	2384.2	1.000	1.082
333.2	0.971	0.986	0.986	2442.9	2464.8	1.000	1.089

Table 25. PTxy data for the 1-butene + methyl tertiary-butyl ether system.

Temp., K	Mole Fraction 1-Butene			Pressure, kPa		Activity Coefficients Calculated	
	Measured Liquid	Measured Vapor	Calculated Vapor	Measured	Calculated	1-Butene	MTBE
293.2	0.0173	0.132	0.147	32.7	29.6	1.050	1.000
293.2	0.126	0.557	0.579	55.1	54.1	1.039	1.001
293.2	0.492	0.889	0.898	137.9	136.6	1.013	1.012
293.2	0.977	0.997	0.997	241.3	249.0	1.000	1.049
313.2	0.0159	0.100	0.103	63.8	65.9	1.030	1.000
313.2	0.485	0.855	0.863	241.3	245.8	1.008	1.007
313.2	0.757	0.949	0.953	356.7	354.9	1.002	1.017
313.2	0.978	0.996	0.996	442.9	447.6	1.000	1.029
333.2	0.0494	0.238	0.226	146.5	149.8	1.028	1.000
333.2	0.148	0.502	0.490	212.0	208.4	1.022	1.000
333.2	0.476	0.830	0.827	403.2	408.9	1.008	1.007
333.2	0.757	0.938	0.940	592.8	590.9	1.002	1.017
333.2	0.977	0.995	0.995	737.6	743.4	1.000	1.029
353.2	0.0492	0.196	0.192	261.9	259.4	1.018	1.000
353.2	0.138	0.452	0.420	341.2	336.3	1.015	1.000
353.2	1.454	0.786	0.781	647.9	622.2	1.006	1.004
353.2	0.732	0.919	0.918	920.2	894.9	1.001	1.011
353.2	0.986	0.993	0.994	1168.4	1157.2	1.000	1.019

Table 26. PTxy data for the acetaldehyde + 1-pentene system.

Temp., K	Mole Fraction Acetaldehyde			Pressure, kPa		Activity Coefficients Calculated	
	Measured Liquid	Measured Vapor	Calculated Vapor	Measured	Calculated	Acetaldehyde	1-Pentene
313.2	0.0228	0.0990	0.0991	156.8	154.4	3.697	1.001
313.2	0.141	0.334	0.348	204.0	195.7	2.634	1.031
313.2	0.409	0.506	0.522	225.7	225.5	1.561	1.253
313.2	0.897	0.793	0.797	218.9	210.1	1.016	2.866
313.2	0.973	0.929	0.929	198.2	191.8	1.001	3.542
333.2	0.0199	0.0803	0.0807	277.4	276.9	3.376	1.001
333.2	0.138	0.307	0.328	327.4	349.2	2.460	1.028
333.2	0.293	0.428	0.458	394.3	391.1	1.795	1.120
333.2	0.404	0.510	0.514	401.5	405.0	1.509	1.228
333.2	0.824	0.736	0.735	398.1	396.9	1.039	2.229
333.2	0.993	0.984	0.982	344.7	340.6	1.000	3.271
353.2	0.0181	0.0640	0.0642	468.7	460.8	2.878	1.000
353.2	0.132	0.285	0.298	565.2	569.6	2.225	1.028
353.2	0.397	0.497	0.506	658.3	670.5	1.451	1.189
353.2	0.822	0.754	0.751	661.7	665.9	1.034	2.017
353.2	0.993	0.986	0.985	594.5	584.7	1.000	2.820
373.2	0.0176	0.0552	0.0558	734.1	727.6	2.532	1.000
373.2	0.294	0.406	0.434	971.9	1012.8	1.572	1.090
373.2	0.524	0.766	0.768	1034.0	1058.0	1.029	1.856
373.2	0.994	0.988	0.988	942.6	948.2	1.000	2.459

Table 27. PTxy data for the n-hexane + ortho-chlorotoluene system.

Temp., K	Mole Fraction Acetaldehyde Measured Liquid	Measured Vapor	Calculated Vapor	Pressure, kPa Measured	Calculated	Activity Coefficients Calculated n-Hexane	OCT
393.2	0.0674	0.509	0.505	63.8	65.2	1.354	1.001
393.2	0.287	0.828	0.828	148.2	154.0	1.201	1.028
393.2	0.561	0.919	0.926	248.1	248.1	1.075	1.115
393.2	0.773	0.963	0.965	324.0	317.6	1.020	1.239
393.2	0.824	0.972	0.974	337.8	334.9	1.012	1.279
393.2	0.986	0.998	0.998	393.6	393.4	1.000	1.438
433.2	0.0239	0.194	0.204	127.2	126.9	1.444	1.000
433.2	0.123	0.582	0.574	220.9	222.0	1.354	1.005
433.2	0.546	0.888	0.888	554.9	556.7	1.095	1.122
433.2	0.779	0.950	0.949	737.6	723.9	1.023	1.282
433.2	0.986	0.996	0.997	896.1	888.8	1.000	1.527
473.2	0.0588	0.306	0.317	363.6	366.6	1.476	1.001
473.2	0.275	0.686	0.694	696.2	725.4	1.284	1.030
473.2	0.434	0.797	0.795	937.5	955.9	1.176	1.081
473.2	0.655	0.879	0.881	1313.1	1252.1	1.068	1.214

agreement between the predicted and measured vapor mole fractions indicate good thermodynamic consistency. All of the systems, except the phenol + methyl ethyl ketone system which shows a strong negative deviation from ideality, show positive deviations from ideality. The n-hexane + vinyl acetate and the acetaldehyde + 1-pentene systems formed azeotropes.

Liquid-Liquid Data

The value of the abscissa at breaks between the changing and constant pressure regions in the pressure versus composition curves were assumed to be the compositions of the liquids in equilibrium at 233 and 252 K. Co-workers (2) in the Chemical Engineering Department at Brigham Young University measured the heats of mixing of the carbon disulfide + methanol system at 273 and 293 K. The values of the abscissa at the breaks in the heat versus composition curves were assumed to be the liquid compositions. These compositions are reported in Table 28.

Parameters Used in Data Reduction

The physical data used in the equation of state are given in Table 29 and the parameters used in the Wilson Equation are found in Table 30. Only the carbon disulfide + methanol system, which exhibits liquid-liquid equilibrium, requires a "c" value different from unity. Therefore all systems except the carbon disulfide + methanol are adequately described by the original two parameter Wilson Equation.

Table 28. Liquid-liquid equilibrium for the carbon disulfide + methanol system.

Temperature, K	Mole Fraction Carbon Disulfide Liquid Phase 1	Liquid Phase 2
233.2	0.157	0.997
253.2	0.189	0.977
273.2	0.246[a]	0.942[a]
293.2	0.327[a]	0.889[a]

[a] These values obtained calorimetrically from reference 2.

Table 29. Physical properties used in the modified Redlich-Kwong equation of state.

Compound	T_c, K	P_c, kPa	ω	Reference
Propylene	365.1	4599	0.143	4
1,3-Butadiene	425	4326	0.179	4
n-Hexane	507.3	3029	0.290	4
Ortho-chlorotoluene	656[a]	3910[a]	0.314[a]	
1-Butene	419.6	4022	0.203	4
MTBE	499[a]	3410[a]	0.260[a]	
Methanol	513.2	7952	0.556	4
Pyridine	617	6080	0.297[a]	8
Carbonyl Sulfde	378	6580	0.120	4
Carbon Disulfide	552	7900	0.123	4
Vinyl Acetate	525	4360	0.340	4
Methyl ethyl ketone	535.6	4153	0.329	4
Acetaldehyde	461	5570	0.303	4
1-Pentene	464.7	4052	0.245	4
Phenol	694.2	6129	0.440	4

[a]No value available; estimated by methods of reference 4.

Table 30. Values of Wilson equation parameters used in data smoothing.

System	Temperature, K	Λ_{12}	Λ_{21}	c
Carbon Disulfide(1) + Methanol(2)	233.2	0.39	0.026	1.35
	253.2	0.363	0.068	1.30
	273.2	0.35	0.099	1.25
	293.2	0.335	0.118	1.20
Carbonyl Sulfide(1) + Methanol(2)	233.2	0.5	0.03	1
	253.2	0.5	0.03	1
	273.2	0.5	0.03	1
	293.2	0.5	0.03	1
Methanol(1) + Pyridine(2)	353.2	0.99	0.89	1
	373.2	0.98	0.93	1
	393.2	0.97	0.94	1
	413.2	0.96	0.95	1
n-Hexane(1) + Vinyl Acetate(2)	333.2	0.58	0.50	1
	353.2	0.63	0.55	1
	373.2	0.66	0.58	1
	393.2	0.72	0.62	1
Vinyl Acetae(1) + Methyl Ethyl Ketone(2)	333.2	0.98	0.95	1
	353.2	0.98	0.95	1
	373.2	0.98	0.97	1
	393.2	0.98	0.97	1
Methyl Ethyl Ketone(1) + Phenol(2)	393.2	1.9	2.1	1
	413.2	1.7	2.0	1
	433.2	1.6	1.9	1
	453.2	1.5	1.6	1
Propylene(1) + 1,3-Butadiene(2)	273.2	0.95	0.96	1
	293.2	0.95	0.96	1
	313.2	0.95	0.96	1
	333.2	0.95	0.96	1
1-Butene(1) + Methyl Tert-Butyl Ether(2)	293.2	0.98	0.97	1
	313.2	0.99	0.98	1
	333.2	0.99	0.98	1
	353.2	0.99	0.99	1
Acetaldehyde(1) + 1-Pentene(2)	313.2	0.43	0.46	1
	333.2	0.45	0.52	1
	353.2	0.51	0.57	1
	373.2	0.55	0.63	1
n-Hexane(1) + Ortho-Chlorotoluene	353.2	0.77	0.68	1
	393.2	0.90	0.76	1
	433.2	0.92	0.70	1
	473.2	0.95	0.62	1

Reliability of Data

Sample analyses were reproducible to ±0.5% within a sample and to 1% between samples. Chromatographic response factors were reproducible to better than ±.5% and the compositions of the standards used to determine these factors were known to better than ±0.5%.

The measured mole fractions of the vapors and the liquids are believed to be accurate within ±2% for mole fractions greater than 0.10 but are only good to ±5% at very low mole fractions with the exception of the methyl ethyl ketone + phenol system, where due to experimental difficulties the uncertainties are about twice as large. The predicted vapor mole fractions are within these limits for almost all the PTxy data.

The PTx data are not as precise nor accurate as the PTxy data, due to the effect of trace impurities in the samples and the reliance on models to give the mole fractions. The errors are believed to be accurate within ±3%. The liquid compositions of the carbon disulfide + methanol system are believed to be within ±1%. The pressure readings are accurate to about ±1% of the total pressure.

The fugacity coefficients obtained from the modified Redlich-Kwong equation of state decrease in accuracy as the reduced pressure increases. Errors in the mole fractions due to errors in the fugacity coefficients tend to cancel out. However, errors in the equation of state strongly effect the calculated system pressure. At pressures in the vicinity of 1200 kPa, the errors in the calculated pressure are about 3%.

ACKNOWLEDGEMENT

The authors gratefully acknowledge the financial support of the Design Institute for Physical Properties of the American Institute of Chemical Engineers for financial support for this project.

REFERENCES

1. Ronc, M. and G.A. Ratcliff, Can. J. Chem., **54** (4), 326 (1976).

2. McFall, T.A., M.E. Post, J.J. Christensen, and R.M. Izatt, J. Chem. Thermo., **14**, 509 (1982).

3. Boublik, T., V. Fried, and E. Hala, The Vapor Pressures of Pure Substances, Elsevier, New York (1973).

4. Reid, R.C., J.M. Prausnitz, and T.K. Sherwood, The Properties of Gases and Liquids, 3rd ed., pp. 629 to 665, McGraw-Hill, New York (1977).

5. Scatchard, G. and F.G. Satkiewiez, J. Am. Chem. Soc., **86** (2), 130 (1964).

6. Wilson, G.M., J. Am. Chem. Soc., **86** (2), 127 (1964).

7. Van Ness, H.C., S.M. Byers, and R.E. Gibbs, AIChE J., **19** (2), 238 (1973).

8. Perry, J.H. and Cecil Chilton, Chemical Engineers' Handbook, 5th ed., p **3**-105, McGraw-Hill, New York (1973).

VAPOR-LIQUID EQUILIBRIUM MEASUREMENTS

Jonathan L. Owens, C. Jeffrey Brady, James R. Freeman,
W. Vincent Wilding and Grant M. Wilson ■ Wiltec Research Co., Inc., 488 South 500 West, Provo, UT 84601

Vapor-liquid equilibrium measurements have been performed on six binary systems by the PTx method, and infinite dilution activity coefficients and Henry's constants have been determined from gas chromatographic (GC) retention measurements for three of these same systems. The following systems were studied:

1. 1,2-Dichloroethane - Vinyl Chloride
2. 1,1,2-Trichlorethane - Vinyl Chloride
3. Acetonitrile - Vinyl Chloride
4. Acetonitrile - Ethylacetylene
5. Acetonitrile - Vinylacetylene
6. Triethylamine - Methylethylketone

Henry's constants were determined for the three systems involving acetonitrile. The PTx measurements were done along four isotherms between 0 and 100°C for each binary system. The GC analyses were also performed at four temperatures.

The vapor-liquid equilibrium results have been fit to the NRTL equation for each system at each temperature studied, and the parameters are included with the results.

Of the systems studied, triethylamine-methylethylketone exhibited an azeotrope. This occured at all four temperatures studied and at concentrations between 36 and 40 mole % triethylamine in the liquid. None of the other systems exhibited azeotropes.

Vapor-liquid equilibrium (VLE) data are essential for modeling and process development. If the various binary systems of a given process are well characterized then an accurate description of the multicomponent problem can generally be developed. Under the sponsorship of the Design Institute for Physical Property Data of the American Institute of Chemical Engineers six binary systems were studied to provide accurate VLE data for industrial and modeling endeavors.

The PTx method, which was employed to obtain these data, is a simple yet accurate way to measure VLE data. This procedure requires only the overall composition in the cell, the temperature, and the total pressure. Infinite dilution activity coefficients and Henry's constants were also obtained as part of this project for three of the systems. These were calculated from gas chromatography retention times which give reliable results for this type of data.

APPARATUS AND PROCEDURE

PTx Measurements

The PTx method is a fast, efficient method for the study of binary vapor-liquid equilibria. The principal measurements are total pressure versus charge composition at constant temperature and cell volume. The PTx method is attractive because no vapor or liquid samples are required and thus the usual sampling and analytical problems are avoided.

The PTx measurements were made in a 300 cc stainless steel cell (except for the acetonitrile/vinylacetylene system for which a 150 cc cell was used) shown schematically in Figure 1. Two lines connected to the top of the cell were for charging and degassing the cell and for connection to the manometer for pressure measurements. A third line, at the bottom of the cell, was connected to a pressure gauge. For pressures below 5 kPa a McLeod gauge was used to measure the pressure, at intermediate pressures (to 350 kPa) a manometer was used, and at high pressures (above 350 kPa) precision pressure gauges (3D Instruments Inc.) were used. The temperature was measured using a platinum resistance probe (RTD) which had been calibrated against the vapor pressure of water.

The composition range along each isotherm was studied in two parts. The cell was first charged with one of the pure components, and increments of the second component were added from a weighed charging cylinder. After each increment was added, the cell was allowed to reach equilibrium at the desired temperature and the cell pressure was read. Increments of the second component were added in this fashion until the cell contained a charge of about 60 mole % of the second component.

The second part of the composition range was studied in one of two ways depending on the system. For the methylethyl/ketone-triethylamine system, where the vapor pressures of the two components are similar, the second part of the composition range was studied by charging the cell with the second component and adding increments of the first until a composition of 60 mole % of the first component existed in the cell. For the other five binary systems the second part was studied by first charging the cell with the lower volatility component as in the first part, but this time a much smaller amount was charged. Then enough of the second component was added until the cell contained 40 mole % of the second component. Then increments of the second component were added up to a concentration of about 90 mole %. Again after each increment was added and equilibrium was established, the pressure was read.

This method yielded a region of overlap between the two parts of the procedure from 40 mole % to 60 mole % of each component. This provided an internal consistency check on the data. After each addition the cell was degassed by drawing a small weighed amount of vapor into a degassing cylinder.

All four isotherms for each binary system were studied simultaneously. After an increment had been added, the cell was placed in a controlled-temperature bath at the lowest temperature and the contents of the cell were degassed. Once equilibrium was established and the pressure measured, the cell was transferred to a bath controlled to the next higher temperature. This was repeated for the other temperatures, and then the cell was returned to the coolest bath where the next increment was added and the contents were degassed.

The accuracy of the various measurements is as follows: Below 13.8 kPa, the pressures are accurate to within ± 0.007 kPa. Between 13.8 kPa and 300 kPa pressures were accurate to within ± 0.14 kPa. Above 300 kPa the pressures were accurate to within 0.5 % of the measured pressure. Thus all pressures were accurate to 1.0 % or better. Temperatures were measured to within ± 0.05 °C. Weights were measured to within ± 0.01 grams. The purity and source of the chemicals used in this study are shown in Table 1.

The PTx data reduction procedure is outlined in Gillespie et al. (1).

GC Measurements

The Henry's constants and infinite dilution activity coefficients determined from gas chromatographic retention times were measured by means of the apparatus shown in Figure 2. Helium regulated to 240 psia was used as the carrier gas. From the regulator, the carrier gas entered a liquid temperature bath controlled to the desired temperature. In the bath, the stream passed through a presaturation column packed with solvent on Chromosorb-W to ensure the gas was saturated with solvent before entering the main test column. The pressure at the inlet and outlet to the test column was measured and means were provided for injecting solute samples as shown. The solute compounds and nitrogen were mixed together and then pressurized with helium to produce the solute sample mixture which was injected into the chromatograph. After passing through the main column, the stream was throttled to atmospheric pressure before flowing through the thermal conductivity detector.

Measured data included the following:

1. Total moles of solvent on the column,

2. Temperature of the column,

3. Inlet and outlet pressures of the column,

4. Retention times of the various solutes injected into the column, and

5. Carrier gas flow rate and total cumulative flow of carrier gas through the column at each condition.

The amount of solvent on the column was determined at the time the column was packed. The temperature was measured to an accuracy of ± 0.05 °C and the pressure was determined to within ± 0.7 kPa. The total volume of gas passing through the column and flow rates of the carrier gas were determined by means of a wet test meter connected to the outlet of the thermal conductivity detector.

The GC data reduction procedure is outlined in Gillespie et al. (1).

RESULTS

Tables 2 through 25 give the results of PTx measurements on the six binary systems studied. Included in each table are the run 'half' (Each isotherm was studied in two parts as discussed in the procedure section.), the charge composition, the liquid and vapor phase compositions, the measured and calculated total pressure, the activity coefficients, and the relative volatility. At the bottom of each table the NRTL parameters are given.

Tables 2 through 5 give the VLE results for the system of 1,2-dichloroethane and vinyl chloride at 20.00, 47.00, 73.00, and 100.00 °C, respectively. As seen from the values of the activity coefficients, this system does not deviate strongly from ideality.

The measurements for the system of 1,1,2-trichloroethane and vinyl chloride are summarized in Tables 6 through 9. This system was also studied at 20.00, 47.00, 73.00, and 100.00 °C. Because of the similarity of the components of this system, there is not a strong deviation from ideality.

Tables 10 through 13 give the results of the PTx measurements on the acetonitrile/vinyl chloride binary mixture. These data were measured at 20.00, 46.7, 73.33, and 100.00 °C. This system is more nonideal than the previous two systems because of the chemical dissimilarity of the components.

The results for the fourth system studied, acetonitrile/ethylacetylene, are presented in Tables 14 through 17. This system was studied at 0.00, 27.00, 53.00, and 80.00 °C.

Tables 18 through 21 give the results for the acetonitrile/vinylacetylene binary system. This system was studied at 0.00, 27.00, 53.00, and 80.00 °C. As seen in these tables the charge concentrations of vinylacetylene were always less than 50 mole %. The other half of the composition range of this system was not measured due to the explosive nature of vinylacetylene at high concentrations.

Measurements on the triethylamine/methylethylketone system are summarized in tables 22 through 25. The four isotherms studied were 20.00, 47.00, 73.00, and 100.00 °C. At all four temperatures this system forms an azeotrope at about 39 mole % triethylamine in the liquid.

Table 26 presents infinite dilution activity coefficients obtained from the PTx measurements and also, for the three systems involving acetonitrile, from the GC retention measurements. Table 27 gives the Henry's constants and infinite dilution activity coefficients for the three systems studied by gas chromatography. This work was performed at four temperatures for each of the systems: 0.0, 27.0, 53.0, and 80.0 °C.

Tables 28 through 34 give the pure component vapor pressures obtained as part of this study which were used in the data reduction procedure. Along with measured values, literature values are included for comparison. There is some repetition between tables that have a common component, but in some cases slightly different pure component vapor pressures were measured and slightly different temperatures were used, so each system has been listed on a separate table to avoid ambiguity as to which vapor pressures were used.

DISCUSSION

Figures 3 through 8 give the relative volatilities for each of the binary systems studied. The relative volatility is defined as y/x for the more volatile component (y is the vapor mole fraction and x is the liquid mole fraction.) divided by y/x for the less volatile component. In Figures 3 through 7 the relative volatility is seen to decrease with temperature. However, in Figure 8, which shows the triethylamine/methylethylketone system, at low triethylamine concentrations the relative volatility increases with temperature. Then at about 57 mole % triethylamine there is a pivot point above which the relative volatility decreases with temperature. Note also that this system exhibits an azeotrope (where the relative volatility is equal to 1.0).

Figures 9 and 10 show infinite dilution activity coefficients for each of the systems studied, and a comparison is made between values obtained from PTx studies and those obtained from GC measurements for the three systems that were evaluated by both methods. Good agreement is seen between the two methods.

Figure 11 compares infinite dilution activity coefficients for vinyl chloride in 1,2-dichloroethane and in 1,1,2-trichloroethane measured as part of this project to literature values of several authors. Fairly good agreement exists between the measured results and those of Azbel (4), but there are serious discrepancies with the data of the other three authors. Figure 12 compares the infinite dilution relative volatility of ethylacetylene over vinylacetylene calculated from the results of this study to that of Pavlov (7). Very good agreement is seen.

ACKNOWLEDGMENT

We express gratitude to the American Institute of Chemical Engineers for sponsoring this research.

LITERATURE CITED

1. Gillespie, P.C., J.R. Cunningham, and G.M. Wilson, "Total Pressure and Infinite Dilution Vapor Liquid Equilibrium Measurements for the Ethylene Oxide/Water System," AIChE Symposium Series. Experimental Results from the Design Institute for Physical Property Data I. Phase Equilibria, No. 244, Vol. 81, 26 (1985).

2. Reid, R.C., J.M. Prausnitz, and T.K. Sherwood, The Properties of Gases and Liquids, 3rd Ed. McGraw-Hill Book Co., New York, NY (1977).

3. API Research Project 44.

4. Asbel, I. Ya., Panfilov, A. A., Vdovets, A. V., Gavrilchuk, N.M., Khim. Prom. 44(4), 269 (1968).

5. Golubev, Yu. D., Zh. Prikl. Khim. (Leningrad) 44(3), 574 (1971).

6. Synowiec, L., Zielenski, K., Przemysl Chem. 47(6), 363 (1968).

7. Pavlov, S.Yu., Gorshkov, V.A., Zaikina, T.G., Bushin, A.N., Skorikova, V.V., Khim. Prom. 46(11), 810 (1970).

Table 1. Source and Purity of Chemicals.

Chemical	Supplier	Purity [a]
Vinyl Chloride	Ethyl Corporation	Unknown [b]
Acetonitrile	Aldrich (Gold Label)	99+%
1,2 Dichloroethane	Aldrich (Gold Label)	99+%
1, 1, 2 Trichloro-ethane	Aldrich	95%
2 Butanone (MEK)	Aldrich	99+%
Triethylamine	Aldrich	99%
Ethylacetylene	Linde	95% [c]
Vinylacetylene	Synthatron	99% [d]

(a) All chemicals except vinyl acetylene were further purified during charging. This was done by drawing liquid phase into the charging cylinder, then removing vapor from the charging cylinder.

(b) The vinyl chloride was taken from a product stream, and not analyzed. Vapor pressure measurements done as part of this work indicated purity comparable to the other chemicals used.

(c) The specified purity is for the liquid phase. There were apparently significant lights in the supply cylinder so the charging cylinder was filled from the liquid phase and a large amount (approximately one quarter by weight) of vapor was vented off to remove these lights. After venting, the vapor phase was analyzed by GC, and showed 99+ area% purity. The liquid phase was then charged to the cell.

(d) Due to the high reactivity of vinylacetylene the supplier adds 0.05 wt% 2,6 di-tert-butyl-4-methylphenol as an inhibitor and mixes in 50 wt% xylene. The 99% specified was prior to the mixing. For this chemical the supply cylinder was used as the charging cylinder and the vapor phase was added to the cell. GC analysis of the vapor phase showed area % vinyl acetylene greater than 99.5%.

Table 2. 1,2-Dichloroethane - Vinyl Chloride Vapor-Liquid Equilibrium Measurements by PTx Method at 20.00°C.

Run	Mole Percent Dichloroethane			Pressure, kPa		Activity Coefficient		Relative Volatility
Half	Charge	Liquid	Vapor	Meas.	Calc.	DCE	VC	VC/DCE
(a)	0.00	0.00	0.00	344.05	344.05	1.434	1.000	24.93
2	9.37	9.58	0.39	315.43	311.22	1.329	1.004	27.20
2	18.04	18.85	0.78	281.37	281.51	1.248	1.014	29.46
2	27.49	29.17	1.27	250.14	249.79	1.177	1.033	32.04
2	39.28	42.11	2.01	211.74	210.68	1.110	1.067	35.38
2	48.38	51.66	2.74	182.50	181.42	1.073	1.099	37.93
1	50.41	51.07	2.69	185.81	183.24	1.075	1.097	37.77
1	58.88	59.74	3.57	156.10	155.96	1.049	1.131	40.14
2	59.90	63.44	4.04	145.00	143.99	1.040	1.147	41.17
1	70.59	71.59	5.48	116.52	116.73	1.023	1.186	43.48
2	73.47	76.69	6.82	98.94	98.94	1.015	1.212	44.96
1	79.40	80.21	8.10	85.36	86.30	1.011	1.232	46.00
1	88.98	89.41	14.76	52.68	51.61	1.003	1.287	48.78
1	93.83	94.05	23.96	32.96	33.13	1.001	1.317	50.21
1	100.00	100.00	100.00	8.34	8.34	1.000	1.358	0.00

NRTL PARAMETERS A = 0.230 B = 0.071 C = -1.000

(a) Measured separately.

Table 3. 1,2-Dichloroethane - Vinyl Chloride Vapor-Liquid Equilibrium Measurements by PTx Method at 47.00°C.

Run	Mole Percent Dichloroethane			Pressure, kPa		Activity Coefficient		Relative Volatility
Half	Charge	Liquid	Vapor	Meas.	Calc.	DCE	VC	VC/DCE
(a)	0.00	0.00	0.00	730.84	730.84	1.437	1.000	14.49
2	9.55	9.61	0.66	670.16	658.86	1.324	1.004	16.00
2	18.48	19.00	1.32	597.08	593.88	1.239	1.015	17.50
2	28.29	29.59	2.14	525.38	524.12	1.166	1.035	19.21
2	40.49	42.90	3.39	439.19	438.34	1.099	1.070	21.40
2	49.54	52.68	4.61	372.66	374.69	1.064	1.103	23.03
1	50.78	51.34	4.42	380.93	383.55	1.068	1.098	22.80
1	59.36	60.07	5.84	326.81	325.54	1.044	1.130	24.28
2	60.89	65.68	7.05	293.02	287.30	1.031	1.154	25.23
1	71.15	71.97	8.89	242.83	243.25	1.020	1.181	26.31
2	74.14	77.72	11.33	195.81	201.74	1.012	1.208	27.31
1	79.80	80.56	12.97	180.99	180.75	1.009	1.222	27.81
1	89.12	89.66	22.76	111.28	111.22	1.003	1.270	29.41
1	93.87	94.21	34.98	75.29	74.99	1.001	1.295	30.23
1	100.00	100.00	100.00	27.37	27.37	1.000	1.329	0.00

NRTL PARAMETERS A = 0.262 B = 0.022 C = -1.000

(a) Measured separately.

Table 4. 1,2-Dichloroethane - Vinyl Chloride Vapor-Liquid Equilibrium Measurements by PTx Method at 73.00°C

Run Half	Mole Percent Dichloroethane Charge	Liquid	Vapor	Pressure, kPa Meas.	Calc.	Activity Coefficient DCE	VC	Relative Volatility VC/DCE
(a)	0.00	0.00	0.00	1351.36	1351.36	1.434	1.000	8.80
2	9.55	9.63	1.06	1223.81	1212.49	1.306	1.005	9.95
2	18.48	19.36	2.12	1092.81	1085.33	1.213	1.017	11.08
2	28.29	30.52	3.44	950.09	947.61	1.137	1.039	12.34
2	40.49	44.55	5.48	782.55	778.58	1.075	1.075	13.87
2	49.54	54.70	7.48	658.44	655.74	1.045	1.105	14.94
1	50.78	51.26	6.73	692.92	697.51	1.054	1.094	14.58
1	59.36	60.05	8.85	591.22	590.12	1.033	1.122	15.49
2	60.89	66.72	11.04	503.31	507.14	1.021	1.144	16.16
1	71.15	71.99	13.35	439.54	440.73	1.014	1.162	16.68
2	74.14	79.45	18.18	335.77	345.09	1.007	1.188	17.40
1	79.80	80.59	19.17	330.26	330.32	1.006	1.192	17.51
1	89.12	89.68	32.12	211.67	211.07	1.002	1.224	18.37
1	93.87	94.22	46.48	150.51	150.49	1.001	1.241	18.79
1	100.00	100.00	100.00	72.53	72.53	1.000	1.262	0.00

NRTL PARAMETERS A = 0.348 B = −0.132 C = −1.000

(a) Measured separately.

Table 5. 1,2-Dichloroethane - Vinyl Chloride Vapor-Liquid Equilibrium Measurements by PTx Method at 100.00°C.

Run Half	Mole Percent Dichloroethane Charge	Liquid	Vapor	Pressure, kPa Meas.	Calc.	Activity Coefficient DCE	VC	Relative Volatility VC/DCE
(a)	0.00	0.00	0.00	2323.51	2323.51	1.508	1.000	5.20
2	9.55	9.64	1.71	2095.99	2071.44	1.342	1.006	6.13
2	18.48	19.86	3.38	1847.78	1836.87	1.225	1.022	7.08
2	28.29	31.87	5.43	1585.78	1582.13	1.134	1.049	8.15
2	40.49	46.84	8.56	1275.52	1274.82	1.066	1.091	9.41
2	49.54	57.35	11.60	1061.78	1059.45	1.037	1.124	10.24
1	50.78	51.43	9.77	1234.15	1181.11	1.052	1.106	9.78
1	59.36	60.33	12.67	996.28	998.43	1.031	1.134	10.48
2	60.89	69.32	16.84	811.51	812.30	1.017	1.164	11.16
1	71.15	72.35	18.69	751.52	749.26	1.013	1.174	11.38
2	74.14	81.44	26.73	543.30	558.65	1.005	1.204	12.03
1	79.80	80.93	26.13	569.50	569.49	1.006	1.202	11.99
1	89.12	89.91	41.41	379.21	379.92	1.001	1.232	12.61
1	93.87	94.37	56.48	280.82	285.49	1.000	1.247	12.91
1	100.00	100.00	100.00	165.82	165.82	1.000	1.265	0.00

NRTL PARAMETERS A = 0.430 B = −0.250 C = −1.000

(a) Measured separately.

Table 6. 1,1,2-Trichloroethane - Vinyl Chloride Vapor-Liquid Equilibrium Measurements by PTx Method at 20.00°C.

Run	Mole Percent Trichloroethane			Pressure, kPa		Activity Coefficient		Relative Volatility
Half	Charge	Liquid	Vapor	Meas.	Calc.	TCE	VC	VC/TCE
(a)	0.00	0.00	0.00	344.05	344.05	1.453	1.000	67.82
2	8.47	8.58	0.12	315.09	314.09	1.326	1.004	75.40
2	19.54	19.87	0.29	278.55	277.71	1.210	1.019	84.98
2	29.29	30.09	0.46	246.49	245.94	1.138	1.040	93.19
2	38.82	40.25	0.66	211.39	214.31	1.089	1.065	100.93
1	47.85	47.89	0.86	190.43	190.02	1.061	1.087	106.47
2	48.81	50.71	0.94	182.78	180.90	1.053	1.095	108.46
2	57.09	59.00	1.25	157.20	153.50	1.033	1.120	114.12
1	59.08	59.30	1.26	151.82	152.54	1.033	1.121	114.37
1	69.35	69.63	1.86	115.97	116.96	1.016	1.154	121.03
1	79.20	79.66	2.99	79.01	80.89	1.007	1.187	127.20
1	88.34	88.52	5.50	47.23	47.72	1.002	1.217	132.39
1	93.00	93.14	9.13	29.79	29.92	1.001	1.232	135.01
2	100.00	100.00	100.00	2.89	2.89	1.000	1.255	0.00

NRTL PARAMETERS A = 0.384 B = -0.190 C = -1.000

(a) Measured separately.

Table 7. 1,1,2-Trichloroethane - Vinyl Chloride Vapor-Liquid Equilibrium Measurements by PTx Method at 47.00°C.

Run	Mole Percent Trichloroethane			Pressure, kPa		Activity Coefficient		Relative Volatility
Half	Charge	Liquid	Vapor	Meas.	Calc.	TCE	VC	VC/TCE
(a)	0.00	0.00	0.00	730.84	730.84	1.420	1.000	39.11
2	8.56	8.60	0.21	669.48	663.77	1.298	1.004	43.86
2	19.56	20.17	0.50	583.98	581.21	1.187	1.019	49.93
2	29.36	30.80	0.80	510.55	508.76	1.120	1.039	55.15
2	39.03	41.38	1.16	432.64	437.05	1.074	1.064	60.08
1	47.65	48.08	1.45	391.62	391.13	1.053	1.081	63.06
2	49.05	52.21	1.66	368.87	362.57	1.043	1.091	64.85
2	57.06	60.66	2.20	310.95	303.24	1.026	1.114	68.41
1	58.95	59.59	2.12	310.95	310.83	1.028	1.111	67.96
1	69.23	69.96	3.13	236.14	236.38	1.014	1.140	72.17
1	79.27	79.96	4.98	160.58	162.65	1.006	1.168	76.06
1	88.23	88.74	9.03	95.56	96.40	1.002	1.193	79.36
1	92.94	93.28	14.62	62.12	61.58	1.001	1.206	81.02
2	100.00	100.00	100.00	9.43	9.43	1.000	1.225	0.00

NRTL PARAMETERS A = 0.430 B = -0.310 C = -1.000

(a) Measured separately.

Table 8. 1,1,2-Trichloroethane - Vinyl Chloride Vapor-Liquid Equilibrium Measurements by PTx Method at 73.00°C.

Run Half	Mole Percent Trichloroethane Charge	Liquid	Vapor	Pressure, kPa Meas.	Calc.	Activity Coefficient TCE	VC	Relative Volatility VC/TCE
(a)	0.00	0.00	0.00	1351.36	1351.36	1.427	1.000	21.08
2	8.56	8.60	0.39	1230.70	1218.50	1.301	1.004	24.12
2	19.56	20.62	0.91	1058.34	1052.86	1.184	1.020	28.17
2	29.36	31.85	1.45	910.10	907.75	1.114	1.042	31.75
2	39.03	43.02	2.10	760.14	766.39	1.069	1.068	35.18
1	47.65	48.33	2.48	703.26	699.42	1.053	1.082	36.77
2	49.05	54.29	2.99	637.76	623.96	1.038	1.098	38.51
2	57.06	62.87	3.97	528.13	514.42	1.023	1.121	40.97
1	58.95	59.96	3.60	554.33	551.61	1.028	1.113	40.14
1	69.23	70.39	5.23	416.78	417.47	1.013	1.142	43.08
1	79.27	80.35	8.19	284.75	287.52	1.005	1.170	45.83
1	88.23	89.01	14.39	174.02	173.17	1.002	1.195	48.17
1	92.94	93.46	22.45	113.90	113.99	1.001	1.208	49.35
2	100.00	100.00	100.00	26.48	26.48	1.000	1.226	0.00

NRTL PARAMETERS A = 0.480 B = -0.420 C = -1.000

(a) Measured separately.

Table 9. 1,1,2-Trichloroethane - Vinyl Chloride Vapor-Liquid Equilibrium Measurements by PTx Method at 100.00°C.

Run Half	Mole Percent Trichloroethane Charge	Liquid	Vapor	Pressure, kPa Meas.	Calc.	Activity Coefficient TCE	VC	Relative Volatility VC/TCE
(a)	0.00	0.00	0.00	2323.51	2323.51	1.492	1.000	10.25
2	8.56	8.58	0.76	2102.88	2074.13	1.343	1.005	12.28
2	19.56	21.26	1.75	1768.49	1760.21	1.202	1.024	15.16
2	29.36	33.38	2.74	1496.15	1488.82	1.119	1.051	17.82
2	39.03	45.31	3.91	1218.98	1231.95	1.068	1.083	20.36
1	47.65	48.61	4.30	1170.72	1161.98	1.058	1.093	21.07
2	49.05	57.00	5.49	1003.18	984.32	1.037	1.118	22.82
2	57.06	65.60	7.20	821.16	801.86	1.021	1.144	24.60
1	58.95	60.43	6.09	915.62	911.63	1.030	1.128	23.53
1	69.23	70.91	8.67	689.47	689.17	1.014	1.161	25.68
1	79.27	80.82	13.21	472.98	478.27	1.006	1.193	27.69
1	88.23	89.34	22.17	293.02	296.46	1.002	1.220	29.40
1	92.94	93.67	32.84	199.60	203.74	1.001	1.234	30.27
2	100.00	100.00	100.00	68.33	68.33	1.000	1.254	0.00

NRTL PARAMETERS A = 0.530 B = -0.500 C = -1.000

(a) Measured separately.

Table 10. Acetonitrile - Vinyl Chloride Vapor-Liquid Equilibrium

Measurements by PTx Method at 20.00°C.

Run	Mole Percent Acetonitrile			Pressure, kPa		Activity Coefficient		Relative Volatility
Half	Charge	Liquid	Vapor	Meas.	Calc.	ACN	VC	VC/ACN
(a)	0.00	0.00	0.00	341.64	341.64	3.607	1.000	8.81
2	8.70	0.82	8.34	317.09	314.47	2.512	1.018	12.95
2	16.31	19.97	1.37	292.61	292.75	1.910	1.068	17.94
2	27.09	31.79	1.87	277.45	270.07	1.520	1.155	24.52
2	35.55	40.92	2.45	256.07	251.59	1.337	1.243	30.11
2	45.59	53.25	2.89	219.60	222.29	1.181	1.387	38.31
1	55.18	59.76	3.43	199.60	203.74	1.125	1.477	42.96
2	58.17	63.74	3.69	190.64	191.19	1.098	1.536	45.92
1	65.83	70.84	4.51	159.27	165.82	1.060	1.652	51.44
1	75.78	79.75	6.27	133.14	127.97	1.027	1.816	58.83
1	85.66	89.71	11.39	76.46	76.05	1.007	2.030	67.85
1	91.89	94.45	19.02	47.50	47.16	1.002	2.145	72.47
1	96.09	97.70	35.93	26.06	25.51	1.000	2.230	75.78
1	100.00	100.00	100.00	9.31	9.31	1.000	2.293	0.00

NRTL PARAMETERS A = 0.600 B = 0.190 C = -1.000

(a) Measured separately.

Table 11. Acetonitrile - Vinyl Chloride Vapor-Liquid Equilibrium

Measurements by PTx Method at 46.7°C.

Run	Mole Percent Acetonitrile			Pressure, kPa		Activity Coefficient		Relative Volatility
Half	Charge	Liquid	Vapor	Meas.	Calc.	ACN	VC	VC/ACN
(a)	0.00	0.00	0.00	732.23	732.23	3.279	1.000	5.68
2	9.80	9.91	1.34	676.38	672.24	2.357	1.017	8.13
2	19.68	20.48	2.26	622.60	621.36	1.818	1.065	11.13
2	30.81	32.93	3.14	568.13	566.41	1.461	1.152	15.14
2	39.26	42.65	3.85	524.00	521.18	1.292	1.241	18.59
2	50.70	55.75	5.05	443.33	449.95	1.150	1.388	23.70
1	59.26	60.28	5.60	417.82	421.41	1.115	1.448	25.59
2	60.50	66.53	6.56	374.39	377.76	1.078	1.536	28.33
1	70.21	71.44	7.56	330.95	339.64	1.055	1.612	30.59
1	79.09	80.31	10.45	264.76	260.62	1.025	1.765	34.95
1	89.22	90.10	18.44	156.51	157.20	1.006	1.961	40.25
1	94.13	94.68	29.32	102.04	101.49	1.002	2.064	42.93
1	97.55	97.80	49.80	61.16	60.81	1.000	2.139	44.84
1	100.00	100.00	100.00	30.61	30.61	1.000	2.194	0.00

NRTL PARAMETERS A = 0.570 B = 0.180 C = -1.000

(a) Measured separately.

Table 12. Acetonitrile - Vinyl Chloride Vapor-Liquid Equilibrium Measurements by PTx Method at 73.33°C.

Run	Mole Percent Acetonitrile			Pressure, kPa		Activity Coefficient		Relative Volatility
Half	Charge	Liquid	Vapor	Meas.	Calc.	ACN	VC	VC/ACN
(a)	0.00	0.00	0.00	1360.34	1360.34	3.246	1.000	3.55
2	9.80	9.97	2.12	1265.88	1249.82	2.332	1.017	5.11
2	19.68	21.06	3.61	1150.74	1149.77	1.784	1.068	7.13
2	30.81	34.59	5.04	1041.11	1036.35	1.420	1.164	9.95
2	39.26	45.25	6.23	939.76	939.90	1.253	1.265	12.45
2	50.70	59.30	8.29	801.86	788.69	1.120	1.429	16.11
1	59.26	60.94	8.60	755.67	768.70	1.109	1.451	16.57
2	60.50	70.22	10.88	642.59	643.97	1.058	1.584	19.31
1	70.21	72.21	11.53	597.78	614.53	1.050	1.615	19.93
1	79.09	81.02	15.74	475.74	471.26	1.022	1.764	22.84
1	89.22	90.56	26.71	290.96	289.86	1.005	1.950	26.33
1	94.13	94.96	40.16	198.22	196.08	1.001	2.045	28.07
1	97.55	97.92	61.89	129.28	129.00	1.000	2.113	29.30
1	100.00	100.00	100.00	80.05	80.05	1.000	2.163	0.00

NRTL PARAMETERS A = 0.570 B = 0.170 C = -1.000

(a) Measured separately.

Table 13. Acetonitrile - Vinyl Chloride Vapor-Liquid Equilibrium Measurements by PTx Method at 100.00°C.

Run	Mole Percent Acetonitrile			Pressure, KPa		Activity Coefficient		Relative Volatility
Half	Charge	Liquid	Vapor	Meas.	Calc.	ACN	VC	VC/ACN
(a)	0.00	0.00	0.00	2360.08	2360.08	3.158	1.000	2.32
2	9.80	10.04	3.23	2160.13	2168.89	2.289	1.017	3.34
2	19.68	21.96	5.56	1990.52	1975.01	1.735	1.071	4.79
2	30.81	37.25	7.85	1744.38	1740.31	1.365	1.183	6.97
2	39.26	49.20	9.78	1539.60	1538.43	1.205	1.300	8.94
2	50.70	63.95	13.12	1225.89	1241.89	1.089	1.481	11.75
1	59.26	61.81	12.52	1257.61	1289.18	1.102	1.452	11.31
2	60.50	74.43	17.17	979.75	988.23	1.041	1.637	14.03
1	70.21	73.18	16.46	1005.26	1024.36	1.046	1.617	13.84
1	79.09	81.88	22.19	781.18	782.35	1.020	1.764	15.85
1	89.22	91.09	35.82	495.04	494.77	1.005	1.943	18.33
1	94.13	95.27	50.75	355.77	351.70	1.001	2.032	19.54
1	97.55	98.06	71.23	255.45	251.38	1.000	2.096	20.40
1	100.00	99.99	100.00	179.26	179.26	1.000	2.141	0.00

NRTL PARAMETERS A = 0.560 B = 0.170 C = -1.000

(a) Measured separately.

Table 14. Acetonitrile - Ethylacetylene Vapor-Liquid Equilibrium

Measurements by PTx Method at 0.00°C.

Run	Mole Percent Acetonitrile			Pressure, kPa		Activity Coefficient		Relative Volatility
Half	Chrge	Liquid	Vapor	Meas	Calc	ACN	ETAC	ETAC/ACN
(a)	00.00	00.00	00.00	73.36	73.36	3.828	1.000	5.53
2	10.26	10.41	1.34	68.46	67.95	2.534	1.022	8.54
2	19.75	20.26	2.09	64.67	63.98	1.916	1.074	11.88
2	31.22	32.25	2.81	59.50	59.49	1.508	1.169	16.45
2	38.99	39.84	3.28	55.98	56.41	1.350	1.244	19.55
1	39.26	39.61	3.26	57.71	56.51	1.354	1.241	19.45
2	53.12	54.17	4.38	48.26	49.32	1.166	1.414	25.79
1	55.52	56.12	4.58	48.13	48.18	1.149	1.441	26.67
1	71.30	72.17	7.05	36.06	36.58	1.051	1.686	34.20
1	82.02	82.82	10.88	26.20	26.19	1.018	1.882	39.50
1	90.83	91.38	19.44	15.79	15.86	1.004	2.061	43.94
1	94.62	94.98	29.19	11.03	10.92	1.001	2.143	45.87
2	100.00	100.00	100.00	3.34	3.34	1.000	2.265	0.00
1	100.00	100.00	100.00	3.34	3.34	1.000	2.265	0.00

NRTL PARAMETERS A = 0.630 B = 0.160 C = -1.000

(a) Measured separately.

Table 15. Acetonitrile - Ethylacetylene Vapor-Liquid Equilibrium

Measurements by PTx Method at 27.00°C.

Run	Mole Percent Acetonitrile			Pressure, kPa		Activity Coefficient		Relative Volatility
Half	Chrge	Liquid	Vapor	Meas	Calc	ACN	ETAC	ETAC/ACN
(a)	00.00	00.00	00.00	200.39	200.39	3.752	1.000	3.81
1	10.39	10.42	1.95	188.23	186.30	2.500	1.022	5.85
1	20.18	20.37	3.05	178.30	175.50	1.894	1.074	8.13
1	32.03	32.60	4.12	162.99	162.98	1.491	1.169	11.26
1	39.48	40.35	4.80	154.03	154.32	1.336	1.245	13.40
1	39.58	39.64	4.74	157.20	155.16	1.348	1.238	13.20
1	53.57	55.06	6.47	131.28	134.15	1.156	1.419	17.72
1	56.04	56.23	6.63	132.10	132.25	1.146	1.435	18.08
1	72.04	72.34	10.13	99.63	100.62	1.050	1.677	23.19
1	82.70	82.98	15.41	72.39	72.70	1.017	1.869	26.78
1	91.30	91.49	26.53	45.23	45.40	1.004	2.043	29.79
1	94.92	95.05	38.18	32.68	32.50	1.001	2.122	31.09
1	100.00	100.00	100.00	12.93	12.93	1.000	2.239	0.00
1	100.00	100.00	100.00	12.93	12.93	1.000	2.239	0.00

NRTL PARAMETERS A = 0.625 B = 0.155 C = -1.000

(a) Measured separately.

Table 16. Acetonitrile - Ethylacetylene Vapor-Liquid Equilibrium Measurements by PTx Method at 53.00°C.

Run	Mole Percent Acetonitrile			Pressure, kPa		Activity Coefficient		Relative Volatility
Half	Chrge	Liquid	Vapor	Meas	Calc	ACN	ETAC	ETAC/ACN
(a)	00.00	00.00	00.00	442.98	442.98	3.604	1.000	2.79
1	10.39	10.43	2.67	424.02	413.26	2.435	1.021	4.24
1	20.18	20.53	4.22	393.00	389.11	1.855	1.072	5.86
1	32.03	33.10	5.74	360.59	360.11	1.464	1.168	8.12
1	39.48	41.13	6.74	337.15	339.71	1.314	1.245	9.67
1	39.58	39.65	6.55	346.80	343.65	1.337	1.230	9.38
1	53.57	56.35	9.16	284.06	292.06	1.141	1.422	12.80
1	56.04	56.41	9.17	292.20	291.85	1.140	1.423	12.82
1	72.04	72.61	13.91	220.22	221.77	1.047	1.658	16.41
1	82.70	83.24	20.79	161.40	161.51	1.016	1.841	18.93
1	91.30	91.67	34.36	103.73	104.16	1.004	2.006	21.02
1	94.92	95.16	47.30	77.74	77.62	1.001	2.079	21.92
1	100.00	100.00	100.00	37.99	37.99	1.000	2.187	0.00
1	100.00	100.00	100.00	37.99	37.99	1.000	2.187	0.00

NRTL PARAMETERS A = 0.615 B = 0.145 C = -1.000

(a) Measured separately.

Table 17. Acetonitrile - Ethylacetylene Vapor-Liquid Equilibrium Measurements by PTx Method at 80.00°C.

Run	Mole Percent Acetonitrile			Pressure, kPa		Activity Coefficient		Relative Volatility
Half	Chrge	Liquid	Vapor	Meas	Calc	ACN	ETAC	ETAC/ACN
(a)	00.00	00.00	00.00	892.17	892.17	3.287	1.000	2.21
1	10.39	10.45	3.47	841.15	833.22	2.295	1.019	3.24
1	20.18	20.82	5.61	784.62	781.18	1.775	1.068	4.42
1	32.03	34.02	7.80	715.67	715.67	1.414	1.163	6.09
1	39.48	42.55	9.27	664.65	668.96	1.274	1.240	7.25
1	39.58	39.68	8.76	690.16	685.32	1.315	1.213	6.85
1	53.57	58.58	12.87	548.82	561.65	1.116	1.418	9.57
1	56.04	56.58	12.32	575.71	576.77	1.131	1.393	9.27
1	72.04	72.91	18.56	434.02	435.84	1.043	1.612	11.81
1	82.70	83.62	27.30	321.64	318.93	1.015	1.781	13.59
1	91.30	91.91	43.03	214.63	213.19	1.003	1.928	15.03
1	94.92	95.31	56.52	166.64	165.63	1.001	1.993	15.65
1	100.00	100.00	100.00	96.11	96.11	1.000	2.087	0.00
1	100.00	100.00	100.00	96.11	96.11	1.000	2.087	0.00

NRTL PARAMETERS A = 0.589 B = 0.129 C = -1.000

(a) Measured separately.

Table 18. Acetonitrile - Vinylacetylene Vapor-Liquid Equilibrium

Measurements by PTx Method at 0.00°C.

Run Half	Mole Percent Acetonitrile Charge	Liquid	Vapor	Pressure, kPa Meas.	Calc.	Activity Coefficient ACN	VAC	Relative Volatility VAC/AC
(a)	0.00	0.00	0.00	82.60	82.60	2.102	1.000	11.29
1	51.76	52.12	3.95	48.40	48.72	1.079	1.196	26.49
1	61.99	62.52	5.47	41.16	40.53	1.041	1.254	28.84
1	71.94	72.83	7.98	31.10	31.50	1.019	1.313	30.91
1	80.05	80.49	11.32	23.72	24.19	1.009	1.356	32.30
1	90.38	90.62	22.17	13.93	13.74	1.002	1.412	33.94
1	100.00	100.00	100.00	3.34	3.34	1.000	1.462	0.00

NRTL PARAMETERS A = 0.550 B = -0.210 C = -1.000

(a) From Reid, Sherwood, and Prausnitz, Antoine Equation. Range -73°C to 32°C; values for 53°C and 80°C extrapolated.

Table 19. Acetonitrile - Vinylacetylene Vapor-Liquid Equilibrium

Measurements by PTx Method at 27.00°C.

Run Half	Mole Percent Acetonitrile Charge	Liquid	Vapor	Pressure, kPa Meas.	Calc.	Activity Coefficient ACN	VAC	Relative Volatility VAC/AC
(a)	0.00	0.00	0.00	226.08	226.08	1.927	1.000	8.29
1	52.05	52.21	5.85	132.72	132.24	1.090	1.181	17.58
1	62.44	62.66	8.00	112.45	110.71	1.049	1.243	19.30
1	72.72	72.98	11.43	86.87	87.15	1.023	1.310	20.93
1	80.38	80.63	15.85	67.29	68.04	1.011	1.362	22.09
1	90.55	90.72	29.32	41.02	40.56	1.002	1.433	23.55
1	100.00	100.00	100.00	12.93	12.93	1.000	1.501	0.00

NRTL PARAMETERS A = 0.447 B = -0.043 C = -1.000

(a) From Reid, Sherwood, and Prausnitz, Antoine Equation. Range -73°C to 32°C; values for 53°C and 80°C extrapolated.

Table 20. Acetonitrile - Vinylacetylene Vapor-Liquid Equilibrium Measurements by PTx Method at 53.00°C.

Run Half	Mole Percent Acetonitrile Charge	Liquid	Vapor	Pressure, kPa Meas.	Calc.	Activity Coefficient ACN	VAC	Relative Volatility VAC/AC
(a)	0.00	0.00	0.00	496.97	496.97	2.074	1.000	5.36
1	52.05	52.29	8.00	293.02	295.67	1.088	1.199	12.60
1	62.44	62.88	10.86	249.31	247.74	1.047	1.264	13.91
1	72.72	73.23	15.33	194.36	195.88	1.021	1.331	15.11
1	80.38	80.86	20.94	152.79	154.30	1.010	1.381	15.95
1	90.55	90.87	36.93	96.53	95.57	1.002	1.449	17.00
1	100.00	100.00	100.00	37.99	37.99	1.000	1.510	0.00

NRTL PARAMETERS A = 0.502 B = -0.098 C = -1.000

(a) From Reid, Sherwood, and Prausnitz, Antoine Equation. Range -73°C to 32°C; values for 53°C and 80°C extrapolated.

Table 21. Acetonitrile - Vinylacetylene Vapor-Liquid Equilibrium Measurements by PTx Method at 80.00°C.

Run Half	Mole Percent Acetonitrile Charge	Liquid	Vapor	Pressure, kPa Meas.	Calc.	Activity Coefficient ACN	VAC	Relative Volatility VAC/AC
(a)	0.00	0.00	0.00	979.19	979.19	1.995	1.000	3.91
1	52.05	52.44	10.99	582.60	578.90	1.096	1.193	8.93
1	62.44	63.04	14.64	492.28	487.95	1.052	1.262	9.95
1	72.72	73.43	20.18	389.55	390.00	1.024	1.335	10.93
1	80.38	81.06	26.88	309.92	312.05	1.011	1.393	11.64
1	90.55	91.10	44.88	203.46	201.61	1.002	1.472	12.57
1	100.00	100.00	100.00	96.11	96.11	1.000	1.545	0.00

NRTL PARAMETERS A = 0.450 B = -0.015 C = -1.000

(a) From Reid, Sherwood, and Prausnitz, Antoine Equation. Range -73°C to 32°C; values for 53°C and 80°C extrapolated.

Table 22. Triethylamine - Methylethylketone Vapor-Liquid Equilibrium Measurements by PTx Method at 20.00°C.

Run	Mole Percent TEA			Pressure, kPa		Activity Coefficient		Relative Volatility
Half	Charge	Liquid	Vapor	Meas.	Calc.	TEA	MEK	MEK/TEA
2	0.00	0.00	0.00	9.68	9.68	2.552	1.000	0.53
2	4.98	5.43	9.24	10.40	10.11	2.346	1.002	0.56
2	9.38	10.22	15.76	10.46	10.41	2.185	1.008	0.61
2	17.84	19.49	25.50	10.86	10.79	1.922	1.031	0.71
2	28.10	30.53	34.06	11.09	11.02	1.674	1.080	0.85
2	36.67	39.61	39.64	11.04	11.08	1.508	1.143	1.00
1	45.60	44.92	42.54	11.09	11.06	1.425	1.191	1.10
2	46.07	49.51	44.90	10.96	11.02	1.360	1.242	1.20
2	56.21	60.00	50.10	10.92	10.86	1.234	1.398	1.49
1	61.94	61.47	50.83	10.72	10.82	1.218	1.426	1.54
1	75.97	75.71	58.85	10.17	10.34	1.094	1.808	2.18
1	90.33	90.27	73.37	9.22	9.18	1.017	2.598	3.37
1	95.00	94.99	82.66	8.19	8.47	1.005	3.030	3.98
1	100.00	100.00	100.00	7.33	7.33	1.000	3.658	0.00

NRTL PARAMETERS A = 0.275 B = 0.575 C = -1.000

Table 23. Triethylamine - Methylethylketone Vapor-Liquid Equilibrium Measurements by PTx Method at 47.00°C.

Run	Mole Percent TEA			Pressure, kPa		Activity Coefficient		Relative Volatility
Half	Charge	Liquid	Vapor	Meas.	Calc.	TEA	MEK	MEK/TEA
2	0.00	0.00	0.00	31.10	31.10	2.228	1.000	0.59
2	5.43	5.42	8.29	31.92	32.15	2.067	1.002	0.63
2	10.23	10.21	14.36	32.82	32.87	1.941	1.007	0.68
2	19.50	19.48	23.80	33.99	33.80	1.733	1.028	0.77
2	30.53	30.52	32.52	34.92	34.32	1.534	1.070	0.91
2	39.61	39.61	38.50	34.92	34.37	1.401	1.124	1.05
1	44.92	44.92	41.70	34.47	34.27	1.335	1.165	1.14
2	49.50	49.51	44.36	34.65	34.11	1.283	1.207	1.23
2	60.00	60.01	50.42	33.78	33.49	1.181	1.334	1.48
1	61.46	61.48	51.29	33.51	33.37	1.169	1.356	1.52
1	75.70	75.73	60.87	31.65	31.71	1.071	1.643	2.01
1	90.26	90.29	76.82	27.79	28.26	1.013	2.172	2.81
1	94.98	95.00	85.69	25.92	26.39	1.003	2.434	3.17
1	100.00	100.00	100.00	23.68	23.68	1.000	2.789	0.00

NRTL PARAMETERS A = 0.250 B = 0.480 C = -1.000

Table 24. Triethylamine - Methylethylketone Vapor-Liquid Equilibrium Measurements by PTx Method at 73.00°C.

Run	Mole Percent TEA			Pressure, kPa		Activity Coefficient		Relative Volatility
Half	Charge	Liquid	Vapor	Meas.	Calc.	TEA	MEK	MEK/TEA
2	0.00	0.00	0.00	81.67	81.67	2.002	1.000	0.66
2	5.43	5.41	7.56	83.60	83.77	1.882	1.002	0.70
2	10.23	10.20	13.29	85.77	85.22	1.785	1.006	0.74
2	19.50	19.47	22.58	87.60	87.09	1.622	1.023	0.83
2	30.53	30.52	31.53	88.94	88.03	1.459	1.060	0.95
2	39.61	39.62	37.84	87.77	87.96	1.348	1.106	1.08
1	44.92	44.93	41.27	87.49	87.61	1.292	1.142	1.16
2	49.50	49.52	44.15	87.42	87.13	1.247	1.178	1.24
2	60.00	60.02	50.71	86.25	85.39	1.159	1.287	1.46
1	61.46	61.49	51.66	85.63	85.07	1.149	1.306	1.49
1	75.70	75.76	61.96	80.53	80.67	1.063	1.550	1.92
1	90.26	90.32	78.29	71.84	72.10	1.011	1.986	2.59
1	94.98	95.02	86.88	67.15	67.68	1.003	2.196	2.88
1	100.00	100.00	100.00	61.50	61.50	1.000	2.474	0.00

NRTL PARAMETERS A = 0.200 B = 0.450 C = -1.000

Table 25. Triethylamine - Methylethylketone Vapor-Liquid Equilibrium Measurements by PTx Method at 100.00°C.

Run	Mole Percent TEA			Pressure, kPa		Activity Coefficient		Relative Volatility
Half	Charge	Liquid	Vapor	Meas.	Calc.	TEA	MEK	MEK/TEA
2	0.00	0.00	0.00	187.43	187.43	1.924	1.000	0.68
2	5.43	5.39	7.28	192.22	191.85	1.803	1.002	0.73
2	10.23	10.17	12.83	194.84	194.82	1.708	1.006	0.77
2	19.50	19.45	21.87	197.95	198.48	1.548	1.024	0.86
2	30.53	30.52	30.73	200.15	199.97	1.396	1.060	0.99
2	39.61	39.64	37.12	199.46	199.31	1.294	1.104	1.11
1	44.92	44.93	40.64	198.36	198.24	1.244	1.136	1.19
2	49.50	49.55	43.67	197.46	196.90	1.204	1.170	1.27
2	60.00	60.04	50.67	192.09	192.37	1.129	1.265	1.46
1	61.46	61.52	51.70	192.22	191.56	1.120	1.281	1.49
1	75.70	75.81	62.94	181.05	180.75	1.049	1.480	1.85
1	90.26	90.37	80.03	160.92	161.59	1.008	1.801	2.34
1	94.98	95.05	88.31	152.65	152.54	1.002	1.943	2.54
1	100.00	100.00	100.00	140.72	140.72	1.000	2.121	0.00

NRTL PARAMETERS A = 0.232 B = 0.362 C = -1.000

Table 26. Summary of Derived Infinite Dilution Activity Coefficients for the Six Binary Systems Studied.

Temperature, °C	Vinyl Chloride - 1,2 - Dichloroethane: VC in DCE	Vinyl Chloride - 1,2 - Dichloroethane: DCE in VC	Vinyl Chloride - 1,1,2-Trichloroethane: VC in TCE	Vinyl Chloride - 1,1,2-Trichloroethane: TCE in VC
20	1.36	1.43	1.26	1.45
47	1.33	1.44	1.23	1.42
73	1.26	1.43	1.23	1.43
100	1.27	1.51	1.25	1.49

Temperature, °C	Vinyl Chloride - Acetonitrile: VC in ACN PTx	Vinyl Chloride - Acetonitrile: VC in ACN GC[(a)]	Vinyl Chloride - Acetonitrile: ACN in VC	Triethylamine - Methylethylketone: TEA in MEK	Triethylamine - Methylethylketone: MEK in TEA
20	2.20	2.25	3.61	2.55	3.66
47	2.19	2.25	3.28	2.45	2.66
73	2.16	2.31	3.25	2.00	2.47
100	2.14		3.16	1.92	2.12

Temperature, °C	Ethylacetylene - Acetonitrile: EAC in ACN PTx	Ethylacetylene - Acetonitrile: EAC in ACN GC	Ethylacetylene - Acetonitrile: ACN in EAC	Vinylacetylene - Acetonitrile: VAC in ACN PTx	Vinylacetylene - Acetonitrile: VAC in ACN GC	Vinylacetylene - Acetonitrile: ACN in VAC[(b)]
0	2.26	2.23	3.76	1.46	1.54	2.10
27	2.24	2.25	3.68	1.50	1.58	1.93
53	2.19	2.25	3.53	1.51	1.64	2.07
80	2.11	2.27	3.29	1.54	1.69	2.00

(a) Interpolated from Table 27.

(b) Extrapolated from PTx data in range of 50 to 90 mole % acetonitrile.

Table 27. Henry's Constants and Infinite Dilution Activity Coefficients of Vinyl Chloride, Ethylacetylene, and Vinylacetylene in Acetonitrile Measured by Gas Chromatography.

Temperature °C	Henry's Constant, kPa[a] VC	EAC	VAC	Infinite Dilution Activity Coefficient VC	EAC	VAC
0	367.5	157.9	122.7	15.38	15.38	10.62
27	858.4	428.2	338.5	15.51	15.51	10.89
53	1635	923.2	747.4	15.51	15.51	11.31
80	2904	1822	1463	16.06	15.65	11.65

(a) The Henry's constant given here is the ratio of the partial pressure over the liquid mole fraction at infinite dilution.

Table 28. Pure Component Vapor Pressures for Vinyl Chloride - 1, 2 - Dichloroethane.

Temperature °C	Vapor Pressure, kPa			
	Vinyl Chloride		1, 2 - Dichloroethane	
	Measured[a]	Literature[b]	Measured[a]	Literature[b]
20	344	336	8.34	8.20
47	731	717	27.37	27.44
73	1351	1317	72.53	71.57
100	2324	2248	165.8	164.7

(a) Used in data reduction.

(b) Reid, Prausnitz and Sherwood, The Properties of Gases and Liquids, Harlacher equation.

Table 29. Pure Component Vapor Pressures for Vinyl Chloride - 1, 1, 2 - Trichloroethane.

Temperature °C	Vapor Pressure, kPa			
	Vinyl Chloride		1, 1, 2 - Trichloroethane	
	Measured[a]	Literature[b]	Measured[a]	Literature[c]
20	344	336	2.89	2.45
47	731	717	9.43	9.390
73	1351	1317	26.48	27.0
100	2324	2248	68.33	67.4

(a) Used in data reduction.

(b) Reid, Prausnitz and Sherwood, Harlacher equation.

(c) Reid, Prausnitz and Sherwood, Antoine equation.

Table 30. Pure Component Vapor Pressures for Vinyl Chloride - Acetonitrile.

Temperature °C	Vapor Pressure, kPa			
	Acetonitrile		Vinyl Chloride	
	Measured[a]	Literature[b]	Measured[a]	Literature[b]
20	9.31	9.36	341.6	337
46.7	30.61	29.9	732.2	711
73.33	80.05	77.8	1360	1324
100	179.3	174	2360	2249

(a) Used in data reduction.

(b) Reid, Prausnitz and Sherwood, Harlacher equation.

Table 31. Pure Component Vapor Pressures for Ethylacetylene - Acetonitrile.

Temperature °C	Vapor Pressure, kPa			
	Ethylacetylene		Acetonitrile	
	Measured[a]	Literature[c]	Measured[a]	Literature[b]
0	73.36	73.02	3.34	3.30
27	200.4	201.5	12.93	12.98
53	443.0	446.5	37.99	38.06
80	892.2	887.4	96.11	96.32

(a) Used in data reduction.

(b) API 44, Antoine equation.

(c) Reid, Prausnitz and Sherwood, Harlacher equation.

Table 32. Pure Component Vapor Pressures for Vinylacetylene - Acetonitrile.

Temperature °C	Vapor Pressure, kPa		
	Vinylacetylene	Acetonitrile	
	Literature[b]	Measured[a]	Literature[b]
0	82.60	3.34	3.30
27	226	12.93	12.98
53	497	37.99	38.06
80	979	96.11	96.32

(a) Used in data reduction.

(b) Reid, Prausnitz and Sherwood, Antoine equation, Range: -73.0 to 32°C, 53 and 80° extrapolated. Used for data reduction.

(c) Reid, Prausnitz and Sherwood, Harlacher equation.

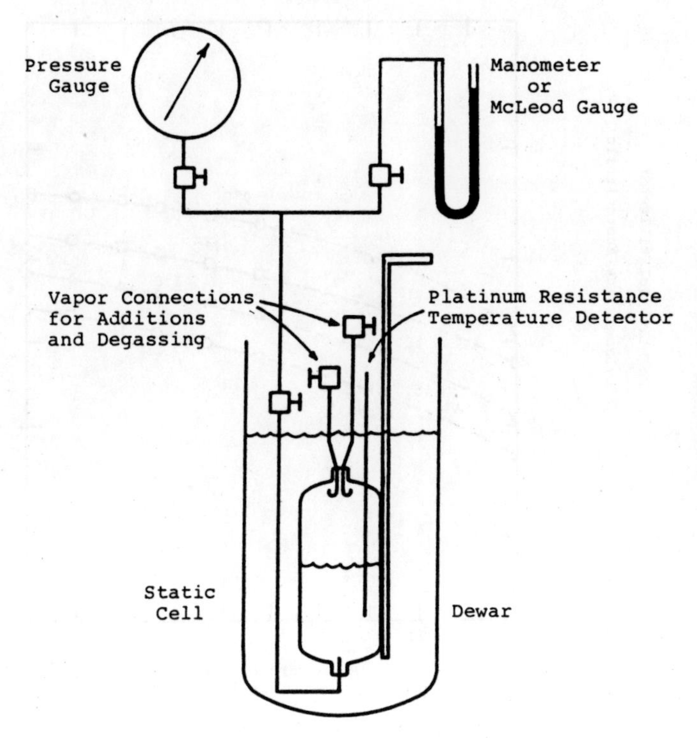

Figure 1. Schematic of agitated static cell used for PTx measurements.

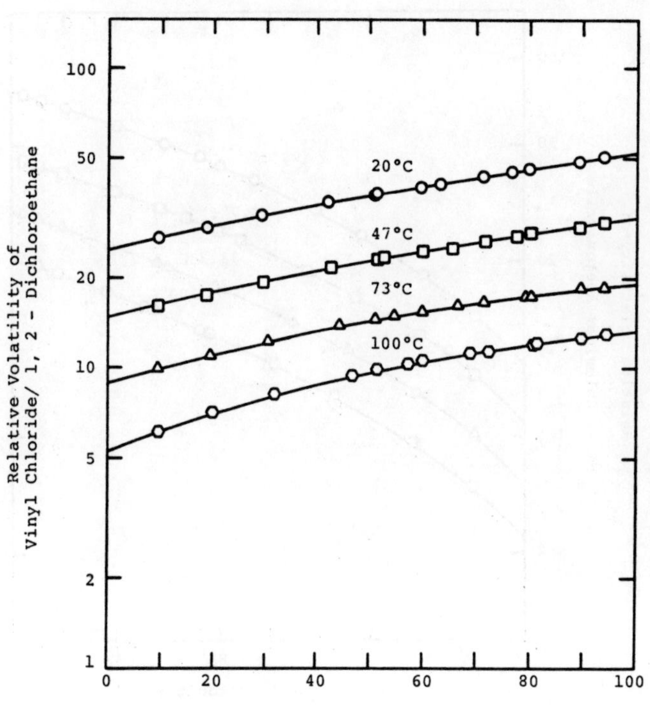

Figure 3. Relative volatility of vinyl chloride over 1,2-dichloroethane based on PTx data.

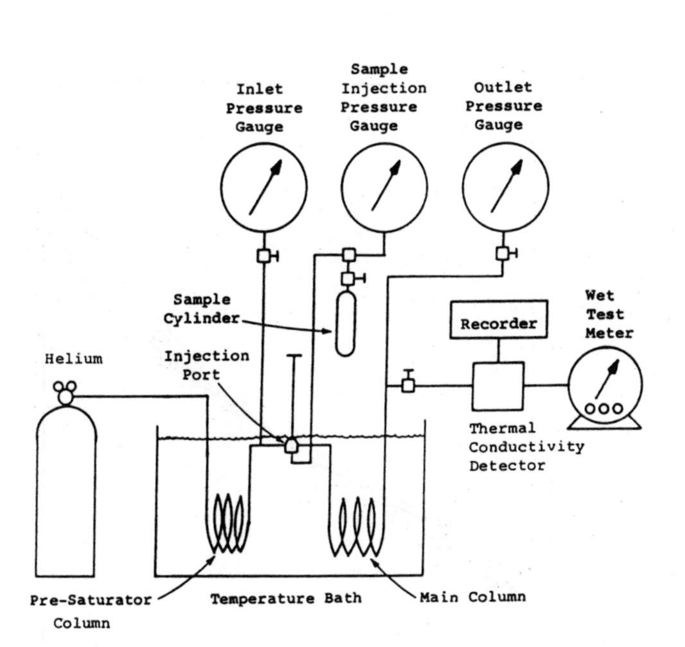

Figure 2. Schematic of gas chromatographic apparatus used for measuring Henry's constants and infinite dilution activity coefficients.

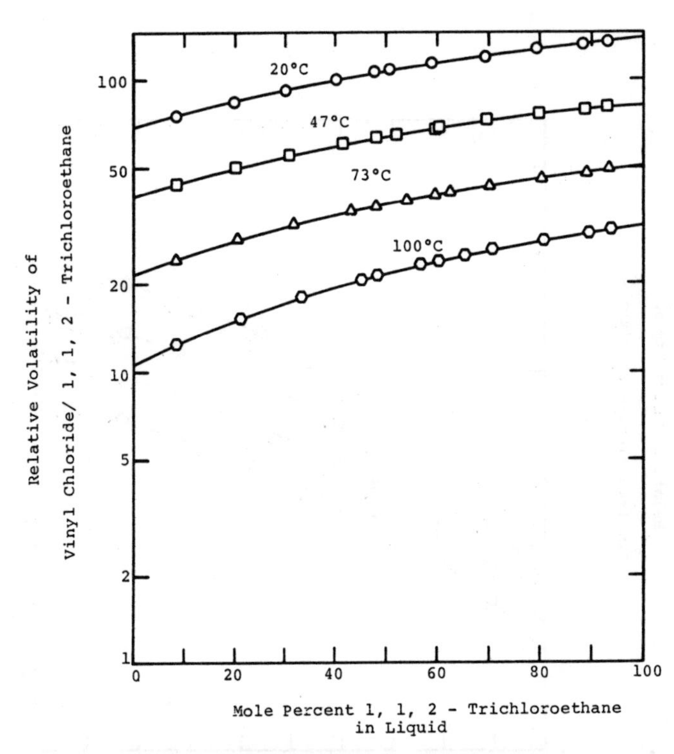

Figure 4. Relative volatility of vinyl chloride over 1,1,2-trichloroethane based on PTx data.

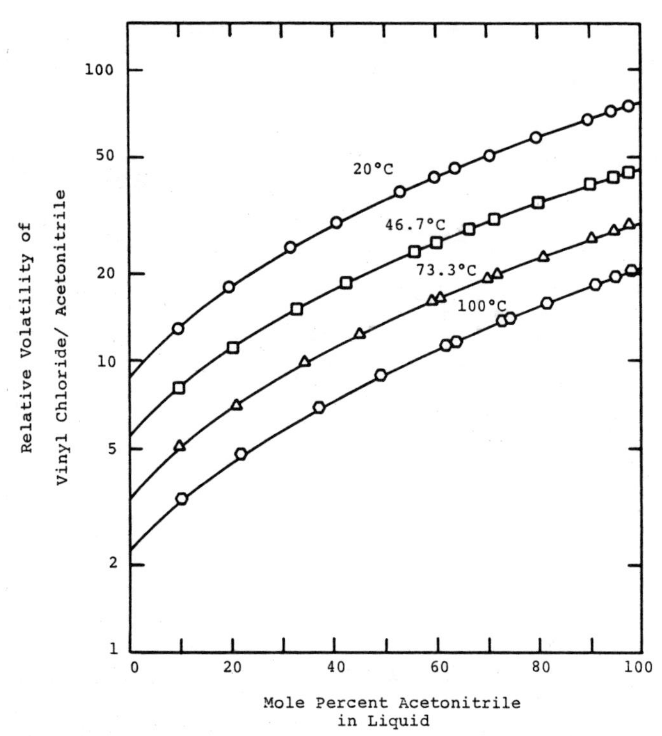

Figure 5. Relative volatility of vinyl chloride over acetonitrile based on PTx data.

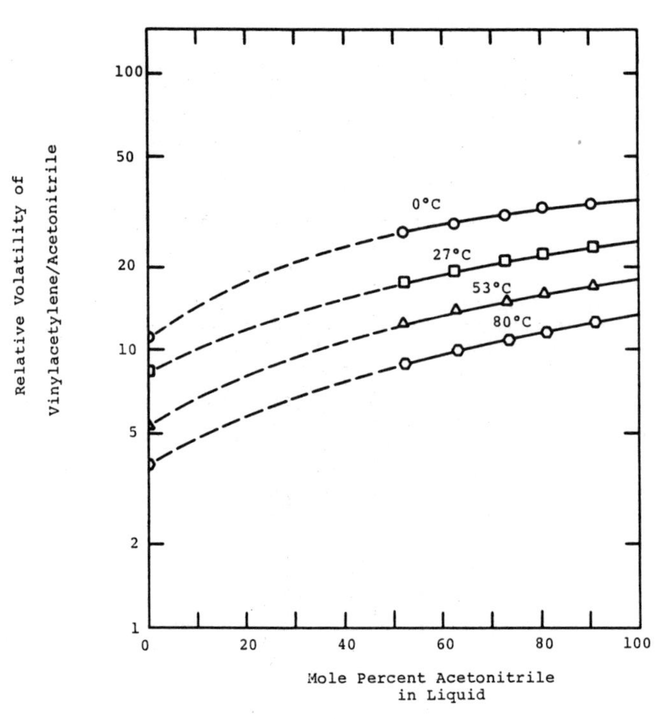

Figure 7. Relative volatility of vinylacetylene over acetonitrile based on PTx data.

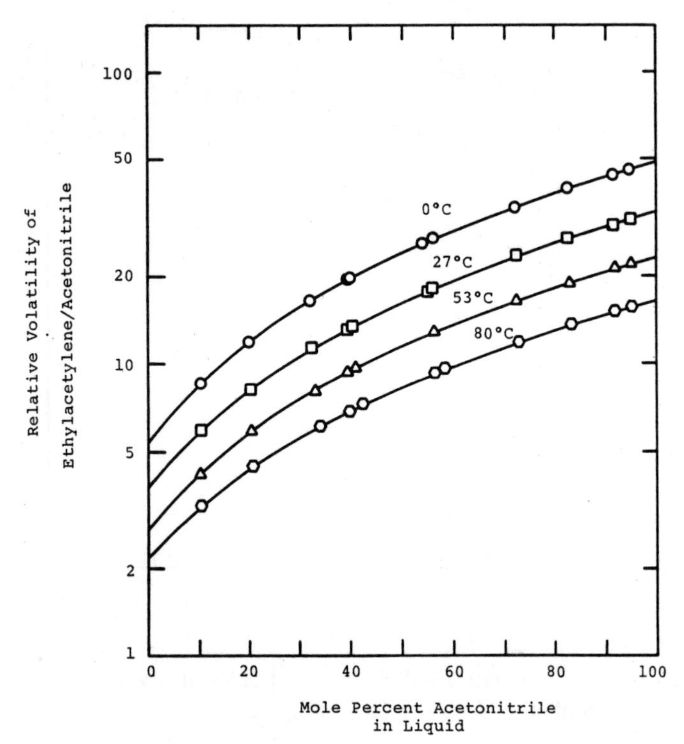

Figure 6. Relative volatility of ethylacetylene over acetonitrile based on PTx data.

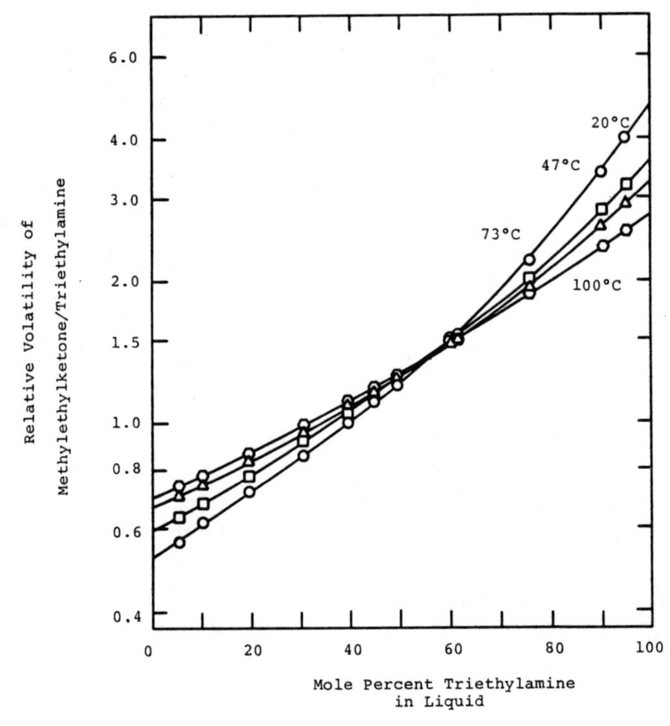

Figure 8. Relative volatility of methylethylketone over triethylamine based on PTx data.

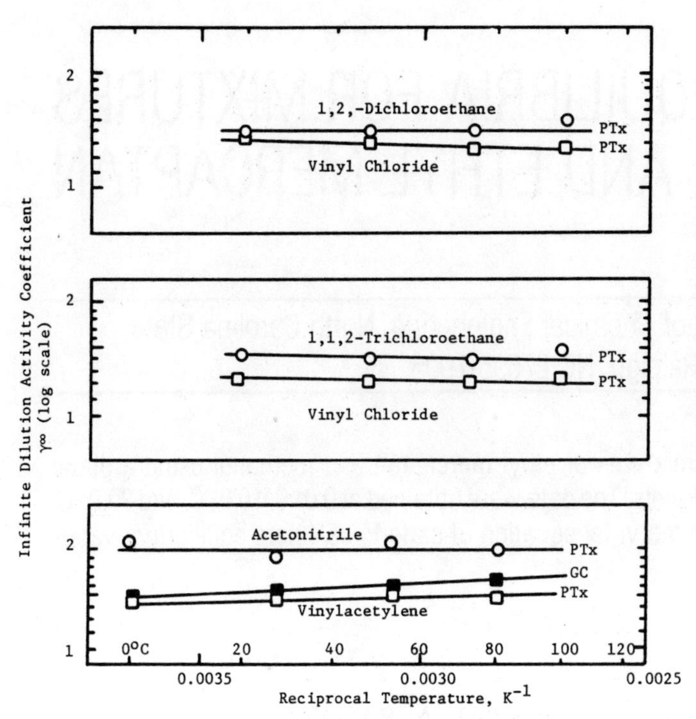

Figure 9. Infinite dilution activity coefficients versus reciprocal temperature for 1,2-dichloroethane—vinyl chloride, 1,1,2-trichlroethane—vinylacetylene and vinyl-acetylene—acetonitrile binary systems.

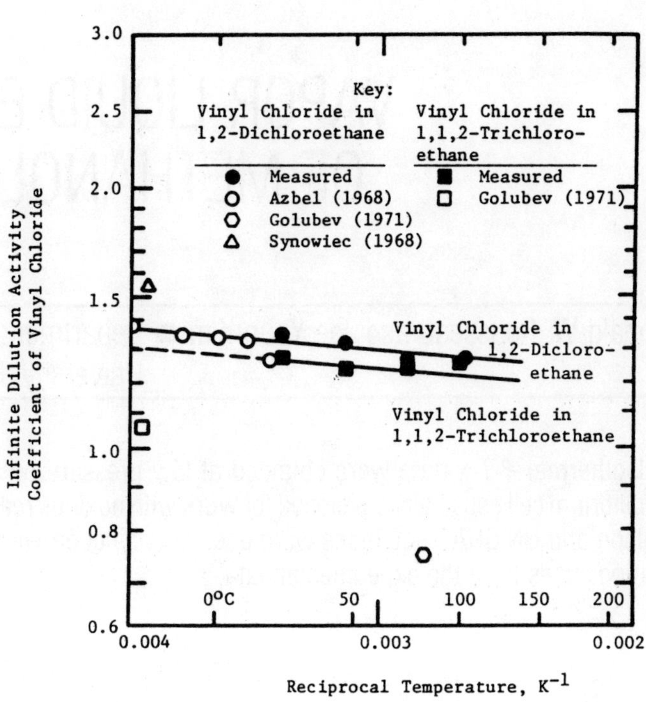

Figure 11. Comparison of vinyl chloride infinite dilution activity coefficients from measured isotherms versus literature data.

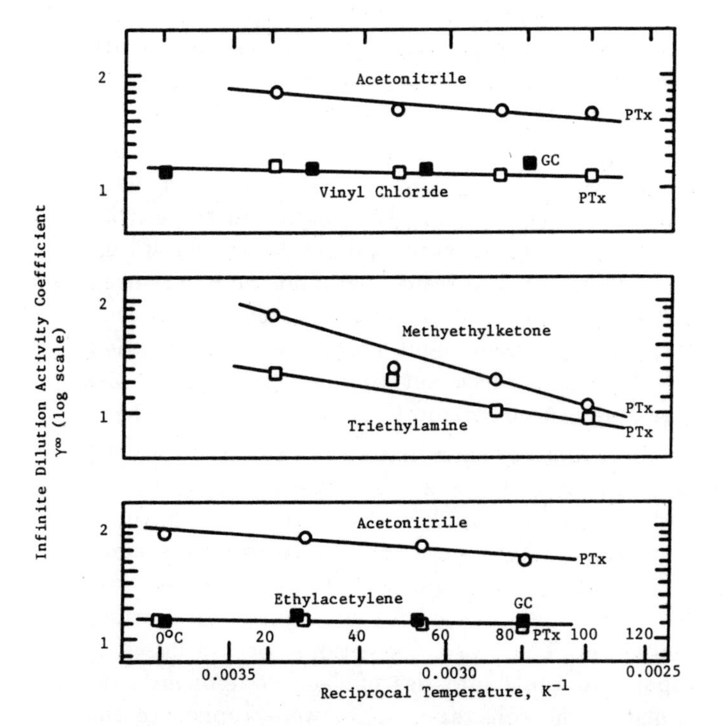

Figure 10. Infinite dilution activity coefficients versus reciprocal absolute temperature for acetonitrile—vinyl chloride, methyethylketone—triethylamine, and aceton-itrile—ethylacetylene binaries.

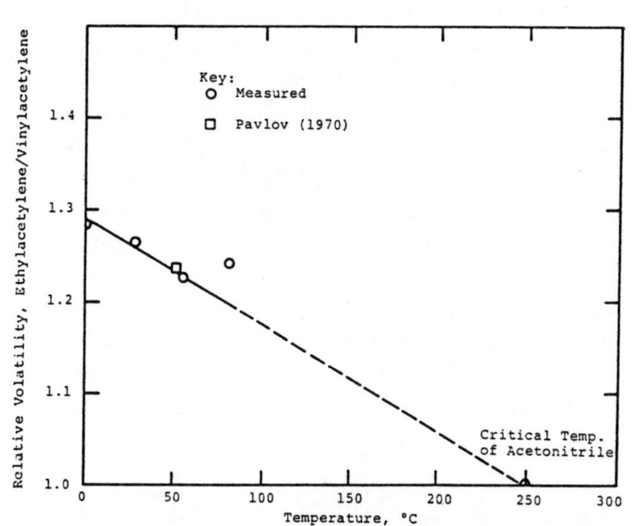

Figure 12. Measured relative volatility of ethylacetylene over vinylacetylene at infinite dilution in acetonitrile and comparison with data point of Pavlov (1970).

VAPOR-LIQUID EQUILIBRIA FOR MIXTURES OF METHANOL AND ETHYL MERCAPTAN

Ronald W. Rousseau and Jae Youn Kim ■ Department of Chemical Engineering, North Carolina State University, Raleigh, NC 27695-7905

Isothermal *P-T-x* data were obtained at low pressure on mixtures of ethyl mercaptan and methanol using a static equilibrium cell especially developed for work with noxious reagents. The data were obtained at 0.0°C, 10.0°C, and 20.0°C. Wilson and UNIQUAC equations were used in conjuction with the virial equation of state to calculate equilibrium vapor compositions from the experimental data.

INTRODUCTION

Conditioning gases produced from coal requires the removal of several sulfur-containing compounds—such as mercaptans—that are deleterious if released to the environment or conveyed to downstream processing units. The removal of these and other undesired species from natural or synthetic gases is accomplished most often by contacting the gas with a solvent that physically or chemically absorbs these components. The solvent is then regenerated through the addition of heat, pressure reduction, or stripping with an inert gas.

Methanol is a solvent commonly used in conditioning gases produced from coal, but little information exits regarding the solubility of mercaptans in this solvent. The data reported here are on mixtures of ethyl mercaptan and methanol, and they complement those presented earlier on mixtures of methyl mercaptan and methanol (1). The temperature range covered by the data is near that at which methanol-based acid gas removal systems operate, and parameters in solution models or equations of state evaluated from the present data will be useful in the design or analysis of these gas conditioning systems.

North Carolina State University, Raleigh, North Carolina. R. W. Rousseau is now with Georgia Institute of Technology, Atlanta, Georgia.

EXPERIMENTAL APPARATUS AND PROCEDURES

The experimental data obtained on mixtures of methanol and ethyl mercaptan included equilibrium pressure, temperature, and liquid composition. The complete experimental apparatus and a description of the procedure involved in its use are given by Kim and Rousseau (1). The chemicals used were commercially obtained reagent grade.

Vapor pressure measurement on these pure substances were performed and compared with literature data. The resulting data and those from the literature for ethyl mercaptan are shown in Figure 1. (Vapor pressures of methanol were measured and reported earlier (1).) The agreement between these measurements and literature data provided an important check on the completeness of the degassing procedure and the purity of the reagents.

The mass of each material introduced into the equilibrium cell gave the overall composition, rather than the liquid composition at equilibrium. In addition, there were minute losses of the components of the vapor phase as the pressure measuring unit was evacuated for subsequent equilibrium conditions. Therefore, corrections for the losses of vapor and the quantities of each component contained in the cell vapor space were applied to the calculation of liquid mole fraction. An iterative procedure, which turns out to be the method used to determine parameters in the liquid solution model, was used to obtain the corrections. The method also involves computation of vapor-phase mole fractions from the model and evaluated parameters. With initial estimates of the liquid compositions, which were obtained from the total moles of each component transferred to equilibrium cell, and the experimental total pressures, estimates of vapor-phase mole fractions and model parameters were obtained. The known volume of the cell vapor space and the number of moles of each component in the vapor phase was calculated and subtracted from the total number of moles added to the cell to obtain revised estimates of the liquid-phase mole fractions. The entire procedure was then repeated, using the revised estimates of the liquid and vapor compositions until successive calculations gave essentially identical compositions.

BINARY EQUILIBRIUM DATA

The experimental data, which are given in Table 1, showed that the system exhibits azeotropes at each temperature investigated and at ethyl mercaptan mole fractions in the range from 0.87 to 0.93. The P-T-x data obtained were reduced using the method of Barker (2), the virial equation of state (truncated after the second virial coefficient) to describe the vapor, and either the Wilson or the UNIQUAC equation to describe the liquid. These calculations use the fundamental relationship

$$\phi_i y_i P = \gamma_i x_i f_i^o \tag{1}$$

Defining ideal solution behavior by Raoult's law and neglecting the Poynting correction factor, Equation 1 becomes

$$\phi_i y_i P = \gamma_i x_i p_i^* \tag{2}$$

The virial equation of state was used to evaluate ϕ_i using the relationship

$$\ln \phi_i = \frac{2}{v} \sum_{j=1}^{2} y_i B_{ij} - \ln z \tag{3}$$

The necessary values of second virial coefficients for this system were estimated using the procedure of Tsonopoulous (3), and they are given in Table 2.

The Wilson equation can be used to calculate activity coefficients (γ_i) using the relationships

$$\ln \gamma_1 = -\ln (x_1 + \Lambda_{12} x_2) + \frac{\Lambda_{12} x_2}{x_1 + \Lambda_{12} x_2} - \frac{\Lambda_{21} x_2}{x_2 + \Lambda_{21} x_1} \tag{4a}$$

$$\ln \gamma_2 = -\ln (x_2 + \Lambda_{21} x_{21}) + \frac{\Lambda_{21} x_1}{x_2 + \Lambda_{21} x_1} - \frac{\Lambda_{12} x_1}{x_1 + \Lambda_{12} x_2} \tag{4b}$$

where

$$\Lambda_{12} = \frac{v_2^L}{v_1^L} \exp \left[\frac{-\Delta\lambda_{12}}{RT} \right]$$

and

$$\Lambda_{21} = \frac{v_1^L}{v_2^L} \exp \left[\frac{-\Delta\lambda_{21}}{RT} \right]$$

The liquid molar volumes, v_1^L and v_2^L, used here are functions of temperature and were estimated using the procedure suggested by Spencer (4) and the values are given in Table 2.

The UNIQUAC equation is mathematically complex, but it has a better theoretical basis than the Wilson equations. For a binary mixture, the activity coefficients γ_1 and γ_2 are given by

$$\ln \gamma_1 = \ln \frac{\Phi_1}{x_1} + \frac{z q_1}{2} \ln \left(\frac{\Theta_1}{\Phi_1} \right) + \Phi_2 \left[l_1 - \frac{l_2 r_1}{r_2} \right] + q_1' \left[-\ln (\Theta_1' + \Theta_2' \tau_{21}) + \frac{\Theta_2' \tau_{21}}{\Theta_1' + \Theta_2' \tau_{21}} - \frac{\Theta_2' \tau_{12}}{\Theta_2' + \Theta_1' \tau_{12}} \right] \tag{5a}$$

$$\ln \gamma_2 = \ln \frac{\Phi_2}{x_2} + \frac{zq_2}{2} \ln \left(\frac{\Theta_2}{\Phi_2}\right) + \Phi_1 \left[l_2 - \frac{l_1 r_2}{r_1}\right] + q_2' \left[-\ln (\Theta_2' + \Theta_1' \tau_{12}) + \frac{\Theta_1' \tau_{12}}{\Theta_2' + \Theta_1' \tau_{12}} - \frac{\Theta_1' \tau_{21}}{\Theta_1' + \Theta_2' \tau_{21}}\right] \quad (5b)$$

where

$$l_1 = \frac{z(r_1 - q_1)}{2} - (r_1 - 1)$$

and

$$l_2 = \frac{z(r_2 - q_2)}{2} - (r_2 - 1)$$

$$\Phi_1 = \frac{x_1 r_1}{x_1 r_1 + x_2 r_2} \quad ; \quad \Phi_2 = \frac{x_2 r_2}{x_1 r_1 + x_2 r_2}$$

$$\Theta_1 = \frac{x_1 q_1}{x_1 q_1 + x_2 q_2} \quad ; \quad \Theta_2 = \frac{x_2 q_2}{x_1 q_1 + x_2 q_2}$$

$$\Theta_1' = \frac{x_1 q_1'}{x_1 q_1' + x_2 q_2'} \quad ; \quad \Theta_2' = \frac{x_2 q_2'}{x_1 q_1' + x_2 q_2'}$$

$$\tau_{12} = \exp\left[\frac{-\Delta u_{12}}{RT}\right] \quad ; \quad \tau_{21} = \exp\left[\frac{-\Delta u_{21}}{RT}\right]$$

and

z = Coordination number = 10

r_1, r_2 = Structural size parameters

q_1, q_2 = Structural area parameters

q_1', q_2' = Modified structural area parameters

The structure size parameter, r, and structure area parameters, q and q' are pure-component physical properties; in the present work q was set equal to q', and the appropriate values of r and q are given in Table 3.

The calculated total pressures and predicted vapor-phase mole fractions for this system are summarized in Tables 4, 5 and 6, and the corresponding P-x-y diagrams are shown in Figure 2. The evaluated parameters for the Wilson and UNIQUAC equations are listed in Table 7. The Wilson equations provided a somewhat better fit to the data, although this may be due to the need to estimate the molecular properties r and q in the UNIQUAC equations.

CONCLUSIONS

Isothermal P-T-x data for ethyl mercaptan-methanol mixtures were obtained, from which thermodynamic models were used to predict the vapor-phase mole fractions. These models use the two-term virial equation in pressure to describe the vapor phase, the Wilson or UNIQUAC equation to express activity coefficients, and pure liquids at the system temperature and pressure as reference states. The parameters of these equations were evaluated using a non-linear parameter search program.

ACKNOWLEDGMENT

Support of this research by the Design Institute for Physical Property Data of the American Institute of Chemical Engineers under Project 805C/81 is acknowledged gratefully.

LITERATURE CITED

1. Kim, J. Y., and R. W. Rousseau, *AIChE Symp. Ser. No. 255,* **81**, 79(1985).
2. Barker, J. A., *Austral. J. Chem.*. **6**, 207 (1953).
3. Tsonopoulous, C., *AIChE J.*. **20,** 263(1974).
4. Spencer, C. F., *J. Chem. Eng. Data,* **17,** 236(1972).
5. Prausnitz, J. M., Personal Communication, 1982.
6. Prausnitz, J. M., T. F. Anderson, E. A. Grens, C. A. Eckert, R. Hsieh and J. P. O'Connell, *Computer Calculations for Multicomponent Vapor-Liquid and Liquid-Liquid Equilibria,* Prentice-Hall, Englewood Cliffs, 1980.

Table 1. **Experimental Vapor-Liquid Equilibrium Data on Mixtures of Ethyl Mercaptan (Component 1) and Methanol**

0.0°C		10.0°C		20.0°C	
x_1	P (mm Hg)	x_1	P (mm Hg)	x_1	P (mm Hg)
0.00000	30.3	0.00000	55.2	0.00000	96.8
0.09545	102.7	0.09516	164.5	0.09479	258.4
0.23100	143.1	0.23032	233.0	0.22942	361.3
0.38959	173.1	0.38941	269.6	0.38872	415.9
0.43151	178.0	0.43126	277.6	0.43093	423.1
0.60962	185.1	0.60954	291.7	0.60944	444.3
0.74778	191.1	0.74763	299.2	0.74748	454.1
0.83845	192.7	0.83846	300.6	0.83849	456.3
0.86930	193.2	0.86929	301.7	0.86932	457.9
0.89929	193.6	0.89931	302.6	0.89938	459.4
0.92133	194.1	0.92137	301.1	0.92147	459.2
0.96954	191.8	0.96960	298.6	0.96970	449.3
1.00000	188.4	1.00000	290.0	1.00000	433.7

Table 2. Second Virial Coefficients (3) and Liquid Molar Volumes (cm^3 /mol) (4)

Component	Property	Temperature (K) 273.15	283.15	293.15
C_2H_5SH	B_{ii}	-1481.99	-1297.06	-1147.52
	v_i	73.75	74.74	75.77
CH_3OH	B_{jj}	-3953.73	-3015.17	-2339.55
	v_j	40.18	40.74	41.33
$C_2H_5SH-CH_3OH$	B_{ij}	-1208.44	-1003.11	-848.89

Table 3. Molecular Properties for UNIQUAC Equation

Component	r	q	Source
C_2H_5SH	2.6154	1.64056	Prausnitz (5)
CH_3OH	1.43	1.43	Prausnitz *et al.* (6)

Table 4. Calculated and Experimental Equilibrium Data on Mixtures of Ethyl Mercaptan (1) and Methanol (2) at 0.0°C

		Wilson Eq.			UNIQUAC Eq.		
x_1	P_{exp}	P_{cal}	ΔP	y_1	P_{cal}	ΔP	y_1
0.09545	102.7	99.49	3.21	0.71832	96.70	6.00	0.71102
0.23100	143.1	147.72	-4.62	0.82688	147.76	-4.66	0.82867
0.38959	173.1	172.52	0.58	0.86359	174.45	-1.35	0.86711
0.43151	178.0	176.30	1.70	0.86905	178.22	-0.22	0.87234
0.60962	185.1	186.14	-1.04	0.88501	186.59	-1.49	0.88542
0.74778	191.1	190.25	0.85	0.89513	189.38	1.72	0.89244
0.83845	192.7	192.36	0.29	0.90474	191.18	1.47	0.90139
0.86930	193.2	192.98	0.22	0.90989	191.82	1.38	0.90713
0.89929	193.6	193.47	0.13	0.91691	192.33	1.27	0.91540
0.92133	194.1	193.66	0.44	0.92422	192.53	1.57	0.92413
0.96954	191.8	192.55	-0.75	0.95435	191.54	0.26	0.95784
RMS			1.85			2.58	

Table 5. Calculated and Experimental Equilibrium Data on Mixtures of Ethyl Mercaptan (1) and Methanol (2) at 10.0°C

		Wilson Eq.			UNIQUAC Eq.		
x_1	P_{exp}	P_{cal}	ΔP	y_1	P_{cal}	ΔP	y_1
0.09516	164.5	162.14	2.36	0.68440	157.99	6.51	0.67707
0.23032	233.0	234.98	-1.98	0.80077	235.38	-2.38	0.80313
0.38941	269.6	271.75	-2.15	0.84111	274.76	-5.16	0.84508
0.43126	277.6	277.26	0.29	0.84712	280.15	-2.60	0.85074
0.60954	291.7	291.56	0.14	0.86502	292.03	-0.33	0.86509
0.74763	299.2	297.55	1.65	0.87691	296.17	3.03	0.87358
0.83846	300.6	300.48	0.12	0.88869	298.73	1.87	0.88495
0.86929	301.7	301.22	0.48	0.89503	299.47	2.23	0.89209
0.89931	302.6	301.63	0.97	0.90365	299.86	2.74	0.90224
0.92137	301.1	301.54	-0.44	0.91254	299.73	1.37	0.91278
0.96960	298.6	298.25	0.35	0.94828	296.53	2.07	0.95231
RMS			1.29			3.20	

$\Delta P = P_{exp} - P_{cal}$

y_1 = predicted vapor mole fractions

RMS = Root Mean Square ΔP

Table 6. Calculated and Experimental Equilibrium Data on Mixtures of Ethyl Mercaptan (1) and Methanol (2) at 20.0°C

		Wilson Eq.			UNIQUAC Eq.		
x_1	P_{exp}	P_{cal}	ΔP	y_1	P_{cal}	ΔP	y_1
0.09479	258.4	256.19	2.21	0.64891	250.18	8.22	0.64160
0.22942	361.3	363.16	-1.86	0.77297	364.12	-2.82	0.77588
0.38872	415.9	416.53	-0.63	0.81717	421.09	-5.19	0.82161
0.43093	423.1	424.50	-1.40	0.82385	428.80	-5.70	0.82782
0.60944	444.3	444.88	-0.58	0.84387	445.48	-1.18	0.84366
0.74748	454.1	453.28	0.82	0.85768	451.38	2.72	0.85375
0.83849	456.3	457.02	-0.72	0.87178	454.63	1.67	0.86763
0.86932	457.9	457.70	0.21	0.87937	455.24	2.66	0.87622
0.89938	459.4	457.65	1.75	0.88967	455.06	4.33	0.88829
0.92147	459.2	456.80	2.40	0.90023	454.06	5.14	0.90068
0.96970	449.3	449.16	0.14	0.94182	446.38	2.92	0.94627
RMS			1.38			4.34	

$\Delta P = P_{exp} - P_{cal}$

y_1 = predicted vapor mole fractions

RMS = Root Mean Square ΔP

Table 7. Wilson and UNIQUAC Parameters Evaluated for Mixtures of Ethyl Mercaptan (1) and Methanol (2)

	Wilson Parameters	
$\Delta\lambda_{12}$	6130.62 atm·cm^3/mol	148.47 cal/mol
$\Delta\lambda_{21}$	60908.14 atm·cm^3/mol	1475.01 cal/mol
	UNIQUAC Parameters	
Δu_{12}	37195.52 atm·cm^3/mol	900.76 cal/mol
Δu_{21}	3648.60 atm·cm^3/mol	88.36 cal/mol

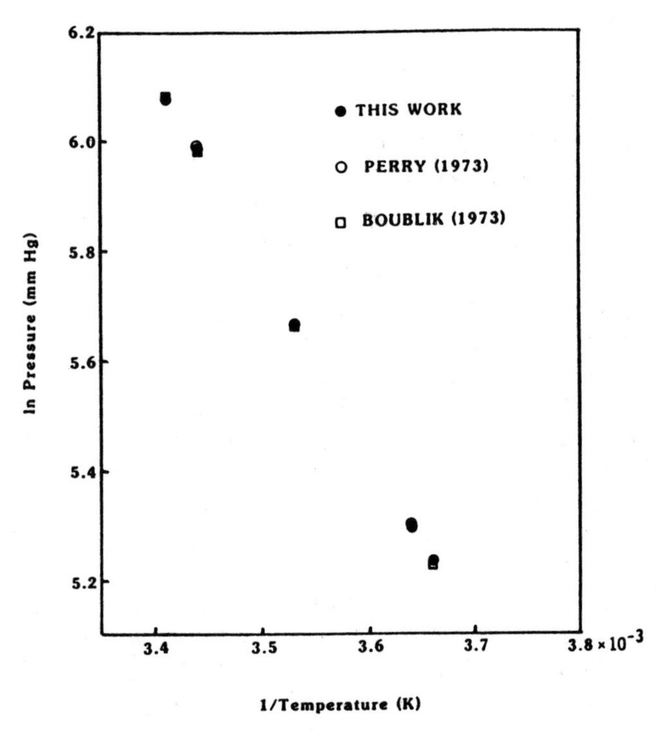

Figure 1. Experimental and literature values of ethyl mercaptan vapor pressures.

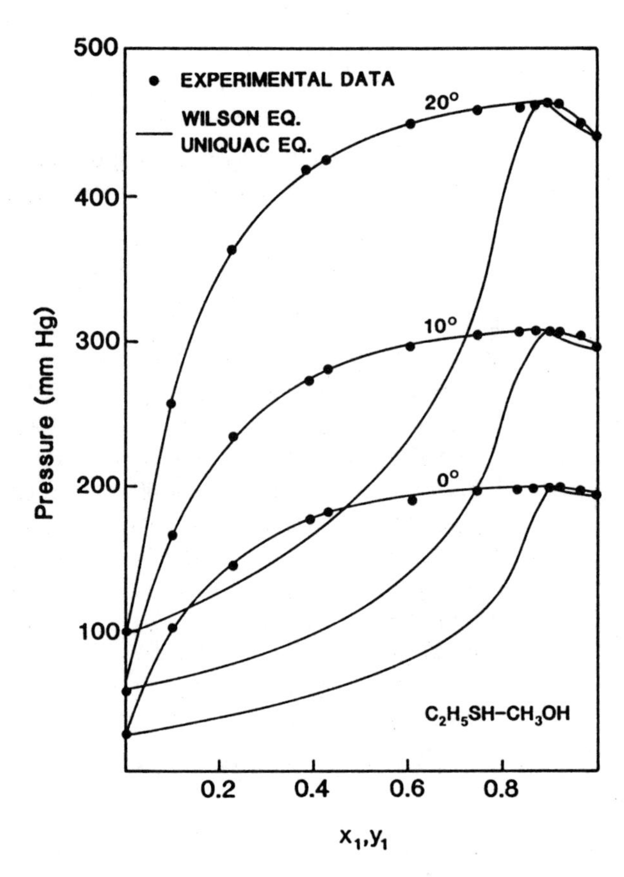

Figure 2. Calculated *P-x-y* plots for ethyl mercaptan (component 1) and methanol using data from this study and the Wilson or UNIQUAC equations.

VAPOR-LIQUID EQUILIBRIUM MEASUREMENTS ON TEN BINARY SYSTEMS OF INDUSTRIAL INTEREST

W. Vincent Wilding, Loren C. Wilson and Grant M. Wilson ■ Wiltec Research Co., Inc., 488 S. 500 W., Provo, UT

Vapor-liquid equilibrium measurements have been made on ten binary systems by the PTx method. Vapor and liquid phase compositions were derived from these data using a modified Redlich-Kwong equation of state to model the vapor phase and the Wilson, NRTL, or Redlich-Kister equation to calculate liquid-phase activity coefficients.

In addition to the PTx measurements, two of the systems, DMF/water and acrylic acid/methyl acrylate, were studied by the PTxy method in which vapor samples were analyzed. These results were compared to the corresponding PTx results. Also, infinite dilution activity coefficients of chloroform in NMP were determined from gas chromatographic retention times.

Project 805 of the Design Institute for Physical Property Data (DIPPR) of the American Institute of Chemical Engineers (AIChE) was established to obtain mixture data, primarily vapor-liquid equilibrium (VLE) data, on systems of industrial interest. These data also contribute new information to predictive techniques in phase equilibria.

Project 805/83 consists of VLE measurements by the PTx method on ten binary systems at two temperatures for each system. Also included are PTxy measurements on two of the ten systems and infinite dilution activity coefficient measurements on one of the systems. The following table lists the systems studied, the type of data taken, and the isotherms at which each system was studied.

Project 805/83

System	Type of Data	Isotherms °C	
1) Benzoic Acid/ Benzonitrile	PTx	150	200
2) Propylene Oxide/ MTBE	PTx	30	90
3) Methyl Iodide/ Acetic Anhydride	PTx	40	140
4) Water/DMF	PTx	30	60
	PTxy	60	
5) Methyl Acrylate/ Acrylic Acid	PTx	40	80
	PTxy	80	
6) Chloroform/NMP	PTx	50	100
	γ^{∞} by GC	50 100 150	
7) Furan/Pyridine	PTx	30	120
8) p-Xylene/NMP	PTx	100	200
9) 1,3-Butadiene/DMF	PTx	40	80
10) Ethylene Glycol/ Diethanolamine	PTx	120	160

APPARATUS AND PROCEDURE

PTx Measurements

The total pressure (PTx) method is a fast efficient method for the determination of binary vapor-liquid equilibria. The required measurements are total pressure versus charge composition at constant temperature and cell volume. The PTx method has the advantage that no vapor or liquid samples are taken, thus eliminating sampling and analyzing problems.

Due to the variety of compounds and conditions involved in this project several different apparatus were used to obtain the PTx data. Systems 2, 3, 7, 8, and 9 were studied using an agitated static cell shown in Figure 1. The apparatus consisted of a 300 cc stainless steel cell in an isothermal bath. The cell was equipped with charging, degassing, and pressure measurement lines. For pressures below approximately 100 kPa, a system of manometers was used to measure the pressure. The lines and first manometer were heated to a temperature higher than the bath to avoid condensation of the vapor outside of the cell. For pressures above 100 kPa, calibrated precision pressure gauges (3D Instruments, Inc.) were used. During a run, the cell was submerged in the constant temperature bath and manually agitated to assure equilibrium. The temperature was measured with a calibrated platinum resistance probe.

Each of the ten binary systems was studied along two isotherms across the entire composition range. Each isotherm was traversed in two parts. First, the cell was charged with a known amount of one component. The cell was then degassed until a repeatable pure component vapor pressure was measured. Increments of the other component were then added to the cell and after the contents of the cell were degassed and allowed to come to equilibrium the pressure was measured. The second part of the procedure was similar to the first except that the second component was charged to the cell initially and increments of the first component were added. The range of mole fractions covered by the two halves were designed to overlap in the mid-composition region to check for consistency between the two parts.

The cell was degassed by weighing an evacuated cylinder or flask before and after a vapor sample was withdrawn. Additions were made using a weighed charging cylinder, syringe, or a positive displacement pump. The amount of material removed by degassing was accounted for in the data reduction procedure.

Systems 1, 4, and 10 were studied in a glass still capable of operation at low pressures. A diagram of this apparatus appears in Figure 2. The temperature of the still was controlled by changing the pressure in the ballast tank. Each isotherm was studied as above except that after the pure components were degassed, no further degassing was performed because the still allowed low boiling contaminants to escape. The contents of the cell were kept in constant reflux by heating the bath to several degrees above the temperature of the cell. The temperature was measured by inserting a calibrated platinum resistance probe into the liquid in the cell. The liquid was kept at saturation conditions by vigorous stirring which aerated the liquid with vapor. The additional surface area and nucleation sites provided by the bubbles helped establish both chemical and thermal equilibrium between the liquid and vapor.

System 5 and the lower isotherm of system 6 were studied in the stirred static apparatus in Figure 3. The glass still was not used for these measurements because the volatility of the lighter component allowed it to escape from the still. The apparatus consisted of a 500 cc glass flask connected to a manometer submerged in an isothermal bath. The submerged manometer served as a null manometer and was read with a cathetometer. This manometer was connected to either a McLeod gauge or another manometer in order to measure the pressure. Each run proceeded as outlined above with the cell being degassed after each addition. Systems 5 and 6 were reactive so run times were kept to a minimum.

The higher isotherm of system 6 (NMP/chloroform) was studied in the apparatus shown in Figure 1 except that the cell volume was 75 cc instead of 300 cc. Originally this system was to be studied at 150 °C, but chloroform proved to be too reactive at this temperature so the system was studied at 100 °C. In order to minimize chemical reaction, each point was a separate charge with the cell being emptied and evacuated between charges. The run time of each point was kept to less than 30 minutes to reduce the effect of decomposition of the chloroform. It appeared that the rate of reaction at 100 °C was slow enough to not affect the measurements.

The PTx data reduction procedure is outlined in Gillespie et al. (1). Critical constants and other pure component data were obtained from Reid, Prausnitz, and Sherwood (2).

PTxy Measurements

PTxy work was performed at the higher isotherm on systems 4 (DMF/water) and 5 (acrylic acid/methyl acrylate) to check for possible vapor interactions. For the DMF/water system, vapor samples were withdrawn into a weighed sample train cooled by dry ice and titrated for water content with Karl-Fisher titrant. The amount of DMF was calculated as the difference between the total sample weight and the amount of water titrated. Since the vapor and charge compositions were known, the liquid composition was calculated by material balance.

The pressure was sufficiently high in the acrylic acid/methyl acrylate system to allow vapor samples to be drawn into a weighed syringe. These samples were titrated with a sodium hydroxide solution to measure the amount of acrylic acid present. The liquid compositions were calculated by material balance. The sodium hydroxide solution was calibrated with both benzoic acid to serve as a primary standard, and with acrylic acid to measure its purity. There was no observable hydrolysis of the methyl acrylate as the titrated solution held its endpoint indefinitely. The PTx apparatus used for both the DMF/water and acrylic acid/methyl acrylate systems were also used for the PTxy measurements.

Infinite Dilution Activity Coefficients

Infinite dilution activity coefficients of chloroform in NMP were measured by gas chromatography. The phase equilibrium of a solute in a liquid solvent on a solid support is directly related to the solute retention time. Measuring retention time, carrier gas flow rate, and column pressure permits the calculation of activity coefficients. This method is particularly suitable for reactive systems since the thermal decomposition affects only the size of the peak, not its retention time.

The apparatus is shown in Figure 4. The columns were 0.75 m long by 6.35 mm o.d. and were packed with 20% NMP on Chromosorb-W. The helium, which was used as the carrier gas, was saturated with NMP in a presaturator bottle and column before reaching the injection port. The carrier gas passed by the injection port and through the main column, and then through a heated line to a thermal conductivity detector. The pressure was measured just before the injection port and just before the needle valve which controls the carrier gas flow rate. The helium flow rate was measured by means of a bubble buret downstream of the detector. The data reduction program required the following data: retention time, the column inlet and outlet pressures, helium flow rate, column temperature, the moles of solvent on the main column, and the vapor pressures of both chloroform and NMP at the column temperature.

The GC data reduction procedure is outlined in Gillespie et al. (1).

RESULTS AND DISCUSSION

The results of the experimental work are summarized in Tables 1 through 23. Each of the tables reporting PTx results gives the run half, the concentration in the feed and in each phase, the measured and calculated pressure, the activity coefficient of each component, and the relative volatility. Consistency between the two halves of each isotherm is demonstrated by the good agreement in the overlap region. The parameters of the activity coefficient equation used to reduce the data are given at the bottom of each table. Three different activity coefficient equations were used: the Wilson, NRTL, and the Redlich-Kister equations. These three equations are given in the appendix.

Tables 1 and 2 present the results for the benzoic acid/benzonitrile system at 150°C and 200°C, respectively. These data were reduced using the Wilson equation with vapor-phase association of benzoic acid taken into account. The association constants (for pressures in mm Hg) used for benzoic acid are 1.1127 x 10-3 at 150°C and 2.0996 x 10-4 at 200°C. The data reduction procedure involving association in the vapor is described in Freeman and Wilson (3). Figure 5 shows the relative volatility of benzonitrile/benzoic acid obtained from the reduced data.

The system of methyl tert-butyl ether and propylene oxide was studied at 30°C and 90°C. These results are summarized in Tables 3 and 4. These data were reduced using the NRTL equation. Figure 6 presents the relative volatility of this system.

Tables 5 and 6 give the vapor-liquid equilibrium measurements for the acetic anhydride/methyl iodide system which were obtained at 40°C and 140°C. The NRTL equation was used to reduce these data. The relative volatility of methyl iodide/acetic anhydride is given in Figure 7.

The DMF/water system was studied by both PTx and PTxy methods. The PTx results, obtained at 30°C and 60°C, are given in Tables 7 and 8. The PTxy results, obtained at 60°C are given in Table 9. The PTx results were reduced using the Redlich-Kister activity coefficient expansion with two parameters. This was required in order to obtain a good fit of the total pressure data. Shown in Figure 8 is the relative volatility obtained from both the PTx and PTxy measurements. There is fair agreement between the PTx and PTxy data with the best agreement at high DMF concentrations. The discrepancy between the two data sets is possibly due to vapor-phase interactions that were not accounted for in the PTx data reduction procedure. If this is the case then the effect will be most pronounced at high pressures (low DMF concentrations). Accounting for vapor interactions would cause the relative volatility calculated from the PTx data to be lowered thus bringing PTx results into closer agreement with the PTxy results.

Tables 10 and 11 present the PTx results for the acrylic acid/methyl acrylate system at 40°C and 80°C. The Wilson equation was used to reduce the data. Vapor-phase association of acrylic acid was taken into account in the data reduction procedure with the association constants 0.55741 at 40°C and 0.03522 at 80°C. PTxy measurements were also performed on this system at 80°C and these results are given in Table 12. Figure 9, which is a plot of the relative volatility for this system, demonstrates excellent agreement between the PTx and PTxy results.

The vapor-liquid equilibrium measurements on the system of NMP and chloroform are given in Tables 13 and 14. These data were obtained at 50°C and 100°C and were reduced using a three-parameter Redlich-Kister expansion. It was originally intended that this system would be studied at 150°C, but because of the rapid reaction of chloroform at that temperature, it was necessary to lower the temperature to 100°C.

Infinite dilution activity coefficients of chloroform in NMP were obtained by gas chromatography at 50°C, 100°C, and 150°C. These results are given in Table 15. Figure 11 is a plot of these data.

In the PTx data reduction procedure for the NMP/chloroform system the Redlich-Kister activity coefficient expansion was used. The first two parameters in this expansion are the infinite dilution activity coefficients of the two components. The values for the infinite dilution activity coefficients for chloroform in NMP obtained by gas chromatography were thus used in the PTx data reduction procedure.

Tables 16 and 17 show the results of PTx measurements on the pyridine/furan system at 30°C and 120°C, respectively. The NRTL equation was used to reduce these data. The relative volatility for this system is shown in Figure 12.

The system composed of NMP and p-xylene was studied at 100°C and 200°C. These data, given in Tables 18 and 19, were reduced using the Wilson equation. Figure 13 presents the relative volatility of p-xylene/NMP.

Tables 20 and 21 give the results of PTx measurements on the DMF/1,3-butadiene system at 40°C and 78.82°C. The Wilson equation was used to reduce these data. Figure 14 gives the relative volatility of 1,3-butadiene/DMF.

The final system studied was diethanolamine and ethylene glycol. This system was studied at 120°C and 160°C and the data were reduced using the Wilson equation. Originally the higher temperature was to be 200°C, but due to decomposition of diethanolamine the temperature was lowered to 160°C. The relative volatility of ethylene glycol/diethanolamine are plotted in Figure 15.

Table 24 is a comparison of measured pure component vapor pressures with literature values. General agreement is seen with the overall average difference being 3.8%. Where discrepancies occur we have measured the vapor pressure several times so that the measured values in Table 24 are reliable.

SUMMARY AND CONCLUSION

Vapor-liquid equilibrium measurements on ten binary systems have been obtained by the PTx method. Two of the systems, DMF/water and acrylic acid/methyl acrylate were also studied using PTxy methods. Infinite dilution activity coefficients were obtained from gas chromatography measurements for the NMP/chloroform system. The results of this work will be useful in design as well as in modeling and correlative efforts.

We express gratitude to Kent Wilson and Mark Swenson for their conscientious experimental work. We also thank the Design Institute for Physical Property Data of the AIChE for sponsoring this research.

<u>REFERENCES</u>

1. Gillespie, P. C., J. R. Cunningham, and G. M. Wilson, "Total Pressure and Infinite Dilution Vapor Liquid Equilibrium Measurements for the Ethylene Oxide/Water System," AIChE Symposium Series. Experimental Results from the Design Institute for Physical Property Data I. Phase Equilibria, No. 244, Vol. 81, 26 (1985).

2. Reid, R. C., J. M. Prausnitz, and T. K. Sherwood, <u>The Properties of Gases and Liquids</u>, 3rd Ed. McGraw-Hill Book Co., New York, NY (1977).

3. Freeman, J. R., and G. M. Wilson, "High Temperature PVT Properties of Acetic Acid/Water Mixtures," AIChE Symposium Series. Experimental Results from the Design Institute for Physical Property Data I. Phase Equilibria, No. 244, Vol. 81, 26 (1985).

4. Wichterle, I., J. Linek, E. Hala, <u>Vapor-liquid Equilibrium Data Bibliography</u>, Elsevier Scientific Publishing Co. (1973)

Table 1. Benzoic Acid/Benzonitrile Vapor-Liquid Equilibrium Measurements by PTx Method at 150.0°C.

Run	Mole Percent BA			Pressure, kPa		Activity Coefficient		Relative Volatility
Half	Feed	Liquid	Vapor	Meas	Calc	BA	BCN	BCN/BA
0	0.00	0.00	0.00	33.67	33.67	1.866	1.000	5.73
2	2.33	2.31	0.40	32.70	33.03	1.787	1.001	5.97
2	4.93	4.91	0.82	31.66	32.33	1.708	1.002	6.24
2	5.77	5.74	0.95	31.57	32.11	1.684	1.003	6.33
2	9.61	9.59	1.55	30.97	31.13	1.586	1.008	6.74
2	19.47	19.53	3.02	28.76	28.72	1.394	1.030	7.79
2	30.22	30.46	4.67	26.27	26.12	1.252	1.067	8.93
2	39.87	40.33	6.36	23.77	23.70	1.165	1.110	9.95
2	50.14	50.90	8.61	20.86	20.93	1.100	1.165	11.00
1	59.92	59.97	11.19	18.24	18.33	1.061	1.218	11.89
2	60.38	61.43	11.70	17.95	17.89	1.056	1.227	12.02
2	70.10	71.40	16.16	14.84	14.68	1.028	1.293	12.96
1	73.12	73.21	17.24	14.00	14.06	1.025	1.305	13.12
2	80.25	81.73	24.37	11.15	10.96	1.011	1.367	13.89
1	84.54	84.64	28.04	9.76	9.83	1.007	1.389	14.14
2	85.60	87.13	32.05	9.37	8.83	1.005	1.409	14.36
1	90.09	90.19	38.61	7.49	7.56	1.003	1.433	14.62
1	94.75	94.84	55.07	5.49	5.55	1.001	1.472	15.01
1	97.84	97.93	75.58	4.08	4.15	1.000	1.498	15.27
2	100.00	100.00	100.00	3.18	3.18	1.000	1.515	-

WILSON PARAMETERS A = 0.480 B = 1.110 C = 1.000

Table 2. Benzoic Acid/Benzonitrile Vapor-Liquid Equilibrium Measurements by PTx Method at 200.0°C.

Run	Mole Percent BA			Pressure, kPa		Activity Coefficient		Relative Volatility
Half	Feed	Liquid	Vapor	Meas	Calc	BA	BCN	BCN/BA
0	0.00	0.00	0.00	126.52	126.52	1.387	1.000	3.95
3	1.65	1.66	0.42	124.04	124.90	1.364	1.000	4.01
3	3.95	4.00	1.01	122.23	122.63	1.334	1.001	4.09
2	5.62	5.62	1.42	119.76	121.09	1.315	1.002	4.15
2	9.97	10.01	2.52	116.28	116.97	1.268	1.005	4.30
2	19.46	19.63	5.03	107.80	108.11	1.187	1.016	4.61
2	29.75	30.14	8.05	98.22	98.47	1.124	1.034	4.93
2	40.08	40.72	11.64	88.22	88.55	1.080	1.058	5.21
2	49.10	49.96	15.50	79.67	79.57	1.052	1.081	5.44
1	59.42	59.47	20.61	70.36	69.92	1.031	1.107	5.65
2	59.60	60.64	21.34	69.28	68.70	1.029	1.110	5.68
1	69.38	69.45	27.96	59.30	59.29	1.016	1.136	5.86
2	71.84	72.93	31.26	55.57	55.45	1.012	1.147	5.92
1	79.82	79.90	39.65	47.27	47.53	1.006	1.169	6.05
1	90.15	90.20	59.68	35.06	35.28	1.001	1.202	6.22
1	94.61	94.64	73.75	29.74	29.78	1.000	1.217	6.29
1	96.74	96.76	82.53	27.03	27.12	1.000	1.224	6.32
2	100.00	100.00	100.00	22.98	22.98	1.000	1.235	-

WILSON PARAMETERS A = 0.551 B = 1.269 C = 1.000

Table 3. Methyl tert-Butyl Ether/Propylene Oxide Vapor-Liquid Equilibrium Measurements by PTx Method at 30.0°C.

Run	Mole Percent MTBE			Pressure, kPa		Activity Coefficient		Relative Volatility
Half	Feed	Liquid	Vapor	Meas	Calc	MTBE	PO	PO/MTBE
2	0.00	0.00	0.00	86.87	86.87	1.344	1.000	1.57
2	2.81	2.82	1.79	85.29	85.95	1.322	1.000	1.60
2	5.56	5.62	3.53	84.39	85.02	1.302	1.001	1.62
2	10.94	11.11	6.94	82.32	83.17	1.264	1.004	1.68
2	18.96	19.34	11.99	80.12	80.32	1.214	1.011	1.76
2	26.90	27.51	17.04	77.50	77.37	1.171	1.022	1.85
2	34.61	35.43	22.08	74.53	74.38	1.134	1.037	1.94
2	42.27	43.29	27.34	71.36	71.27	1.103	1.056	2.03
1	44.87	44.69	28.31	70.95	70.70	1.098	1.060	2.05
2	49.83	51.01	32.88	67.84	68.04	1.076	1.079	2.13
1	52.37	52.28	33.84	67.43	67.49	1.072	1.084	2.14
2	57.29	58.55	38.83	64.88	64.69	1.055	1.107	2.23
1	59.80	59.79	39.88	64.05	64.11	1.051	1.112	2.24
1	67.38	67.44	46.84	60.26	60.42	1.034	1.145	2.35
1	74.82	74.91	54.80	56.54	56.50	1.020	1.183	2.46
1	82.38	82.49	64.58	52.40	52.16	1.010	1.228	2.58
1	89.91	90.01	76.87	47.92	47.39	1.003	1.279	2.71
1	94.93	95.00	87.15	44.68	43.93	1.001	1.318	2.80
1	97.51	97.54	93.30	42.26	42.07	1.000	1.339	2.85
1	100.00	100.00	100.00	40.20	40.20	1.000	1.360	-

NRTL PARAMETERS A = 0.120 B = 0.160 C = -1.000

Table 4. Methyl tert-Butyl Ether/Propylene Oxide Vapor-Liquid Equilibrium Measurements by PTx Method at 90.0°C.

Run	Mole Percent MTBE			Pressure, kPa		Activity Coefficient		Relative Volatility
Half	Feed	Liquid	Vapor	Meas	Calc	MTBE	PO	PO/MTBE
2	0.00	0.00	0.00	544.61	544.61	1.173	1.000	1.51
2	2.81	2.85	1.89	537.86	538.67	1.164	1.000	1.52
2	5.59	5.67	3.77	533.24	532.76	1.156	1.000	1.53
2	11.08	11.21	7.49	521.03	521.01	1.139	1.002	1.56
2	19.29	19.47	13.12	503.11	503.12	1.116	1.005	1.60
2	27.45	27.65	18.85	486.28	484.94	1.095	1.011	1.65
2	35.36	35.56	24.61	466.63	466.84	1.077	1.019	1.69
2	43.23	43.39	30.61	449.26	448.31	1.060	1.030	1.74
1	44.66	44.69	31.65	447.53	445.16	1.058	1.032	1.75
2	50.95	51.06	36.86	429.40	429.49	1.046	1.042	1.79
1	52.24	52.31	37.93	428.30	426.35	1.044	1.045	1.80
2	58.51	58.55	43.45	412.99	410.34	1.033	1.058	1.84
1	59.74	59.85	44.65	408.24	406.93	1.032	1.061	1.85
1	67.38	67.52	52.18	386.86	386.22	1.021	1.080	1.91
1	74.85	75.00	60.43	364.32	364.90	1.013	1.102	1.96
1	82.43	82.58	70.01	341.22	341.98	1.006	1.128	2.03
1	89.97	90.08	81.23	317.57	317.66	1.002	1.159	2.10
1	94.98	95.04	89.93	300.61	300.56	1.001	1.181	2.15
1	97.53	97.57	94.86	291.51	291.50	1.000	1.194	2.17
1	100.00	100.00	100.00	282.54	282.54	1.000	1.207	-

NRTL PARAMETERS A = 0.000 B = 0.160 C = -1.000

Table 5. Acetic Anhydride/Methyl Iodide Vapor-Liquid Equilibrium Measurements by PTx Method at 40.0°C.

Run	Mole Percent AA			Pressure, kPa		Activity Coefficient		Relative Volatility
Half	Feed	Liquid	Vapor	Meas	Calc	AA	MI	MI/AA
2	0.00	0.00	0.00	94.80	94.80	3.518	1.000	16.88
2	2.62	2.65	0.15	92.73	92.49	3.181	1.001	18.70
2	5.19	5.26	0.27	90.46	90.41	2.902	1.005	20.58
2	10.43	10.59	0.48	86.80	86.65	2.456	1.020	24.68
2	20.83	21.15	0.79	80.74	80.39	1.887	1.071	33.80
2	30.97	31.44	1.03	75.50	74.90	1.558	1.146	43.86
2	41.11	41.67	1.28	70.46	69.24	1.350	1.244	55.00
1	45.47	46.02	1.40	66.41	66.63	1.285	1.293	60.09
1	50.04	50.65	1.54	63.19	63.64	1.227	1.350	65.74
2	51.18	51.83	1.58	63.43	62.84	1.214	1.366	67.23
2	56.33	57.06	1.76	58.54	59.05	1.162	1.438	74.01
1	60.03	60.72	1.92	55.65	56.15	1.132	1.494	78.97
1	69.01	69.77	2.45	47.38	47.87	1.074	1.649	92.04
1	80.08	80.74	3.68	34.89	35.07	1.028	1.879	109.76
1	89.83	90.29	6.82	20.61	20.60	1.007	2.126	127.20
1	95.03	95.25	12.77	11.53	11.49	1.002	2.277	137.17
1	97.27	97.38	20.75	7.20	7.20	1.001	2.346	141.65
1	100.00	100.00	100.00	1.53	1.53	1.000	2.438	-

NRTL PARAMETERS A = 0.570 B = 0.250 C = -1.000

Table 6. Acetic Anhydride/Methyl Iodide Vapor-Liquid Equilibrium Measurements by PTx Method at 140.0°C.

Run	Mole Percent AA			Pressure, kPa		Activity Coefficient		Relative Volatility
Half	Feed	Liquid	Vapor	Meas	Calc	AA	MI	MI/AA
2	0.00	0.00	0.00	1161.07	1161.07	2.485	1.000	3.38
2	2.63	2.81	0.78	1134.87	1134.17	2.288	1.001	3.69
2	5.22	5.56	1.45	1109.36	1109.19	2.123	1.004	4.00
2	10.51	11.13	2.63	1062.47	1061.94	1.859	1.017	4.64
2	21.04	21.98	4.51	978.36	977.09	1.519	1.058	5.97
2	31.31	32.31	6.16	894.24	897.81	1.322	1.113	7.28
2	41.55	42.41	7.90	812.89	815.47	1.198	1.180	8.58
1	46.30	46.65	8.74	766.00	778.38	1.160	1.211	9.13
1	50.57	51.03	9.69	732.77	738.08	1.127	1.245	9.71
2	51.74	52.32	10.00	723.25	725.79	1.119	1.255	9.87
2	56.99	57.39	11.33	680.51	675.49	1.089	1.297	10.54
1	60.55	61.24	12.52	633.35	635.10	1.070	1.330	11.04
1	70.50	71.34	16.77	519.03	519.30	1.034	1.422	12.35
1	80.39	81.20	24.06	398.38	391.61	1.013	1.518	13.63
1	90.40	90.94	40.27	253.10	250.95	1.003	1.618	14.89
1	95.15	95.46	57.63	181.33	180.73	1.001	1.667	15.47
1	97.51	97.68	72.76	144.79	145.17	1.000	1.691	15.76
1	100.00	100.00	100.00	107.14	107.14	1.000	1.716	-

NRTL PARAMETERS A = 0.530 B = 0.010 C = -1.000

Table 7. DMF/Water Vapor-Liquid Equilibrium Measurements by PTx Method at 30.0°C.

Run	Mole Percent DMF			Pressure, kPa		Activity Coefficient		Relative Volatility
Half	Feed	Liquid	Vapor	Meas	Calc	DMF	H_2O	H_2O/DMF
2	0.00	0.00	0.00	4.23	4.23	0.496	1.000	12.24
2	2.52	2.53	0.23	4.13	4.13	0.528	0.999	11.49
2	5.00	5.03	0.49	4.03	4.02	0.559	0.997	10.82
2	10.31	10.39	1.20	3.82	3.79	0.626	0.988	9.58
2	20.67	20.84	3.28	3.34	3.31	0.747	0.956	7.77
2	30.28	30.50	6.22	2.89	2.87	0.842	0.918	6.62
1	38.10	38.08	9.34	2.52	2.56	0.902	0.886	5.97
2	40.32	40.57	10.54	2.46	2.46	0.919	0.875	5.79
1	49.98	49.96	15.91	2.10	2.11	0.967	0.839	5.28
2	50.24	50.50	16.26	2.10	2.09	0.969	0.837	5.25
1	60.06	60.05	23.35	1.78	1.78	0.997	0.809	4.94
2	60.13	60.39	23.63	1.79	1.77	0.998	0.808	4.93
1	70.24	70.24	33.06	1.48	1.49	1.009	0.792	4.78
1	80.30	80.31	45.98	1.22	1.22	1.008	0.794	4.79
1	90.08	90.09	64.67	0.97	0.97	1.003	0.819	4.97
1	94.30	94.32	76.52	0.85	0.86	1.001	0.838	5.09
1	96.79	96.80	85.35	0.79	0.79	1.000	0.853	5.19
1	100.00	100.00	100.00	0.69	0.69	1.000	0.875	-

REDLICH-KISTER PARAMETERS A = 0.496 B = 0.875

Table 8. DMF/Water Vapor-Liquid Equilibrium Measurements by PTx Method at 60.0°C.

Run	Mole Percent DMF			Pressure, kPa		Activity Coefficient		Relative Volatility
Half	Feed	Liquid	Vapor	Meas	Calc	DMF	H_2O	H_2O/DMF
2	0.00	0.00	0.00	19.99	19.99	0.814	1.000	6.79
2	2.65	2.65	0.41	19.57	19.53	0.828	1.000	6.68
2	4.91	4.92	0.78	19.21	19.14	0.839	0.999	6.59
2	9.73	9.74	1.66	18.51	18.29	0.861	0.997	6.40
2	19.42	19.44	3.81	16.79	16.56	0.900	0.990	6.08
1	19.73	19.73	3.89	16.57	16.51	0.902	0.989	6.07
2	29.41	29.42	6.70	14.89	14.78	0.933	0.978	5.81
1	30.44	30.44	7.04	14.57	14.60	0.936	0.977	5.78
2	39.56	39.57	10.51	13.01	13.01	0.959	0.965	5.58
1	39.88	39.88	10.64	12.95	12.96	0.959	0.964	5.57
1	49.74	49.75	15.50	11.23	11.28	0.977	0.950	5.39
1	59.93	59.94	22.17	9.59	9.62	0.989	0.936	5.25
1	70.69	70.70	31.90	7.93	7.93	0.996	0.924	5.15
1	80.36	80.37	44.56	6.51	6.47	0.999	0.915	5.09
1	89.44	89.45	62.58	5.14	5.13	1.000	0.911	5.07
1	94.70	94.70	77.93	4.36	4.36	1.000	0.911	5.07
1	97.44	97.44	88.26	3.89	3.96	1.000	0.911	5.07
1	100.00	100.00	100.00	3.59	3.59	1.000	0.912	-

REDLICH-KISTER PARAMETERS A = 0.814 B = 0.912

Table 9. DMF/Water Vapor-Liquid Equilibrium Measurements by PTxy Method at 60.0°C.

Pressure, kPa	Mole Percent DMF Liquid	Mole Percent DMF Vapor	Relative Volatility H_2O/DMF
18.42	10.37	2.23	5.07
14.70	31.26	7.92	5.28
11.27	50.78	17.52	4.85
8.074	69.99	32.03	4.95
4.723	91.71	67.89	5.23

Table 10. Acrylic Acid/Methyl Acrylate Vapor-Liquid Equilibrium Measurements by PTx Method at 40.0°C.

Run	Mole Percent AA			Pressure, kPa		Activity Coefficient		Relative Volatility
Half	Feed	Liquid	Vapor	Meas	Calc	AA	MA	MA/AA
2	0.00	0.00	0.00	22.37	22.37	2.338	1.000	21.51
2	5.65	5.87	0.30	21.39	21.27	1.777	1.008	20.55
2	10.93	11.65	0.63	20.36	20.33	1.493	1.024	20.95
2	21.30	22.91	1.37	18.62	18.54	1.228	1.065	21.32
2	28.40	30.98	2.09	17.06	17.14	1.137	1.096	21.00
2	35.45	39.11	3.06	15.61	15.62	1.082	1.125	20.33
2	46.69	51.17	5.23	13.02	13.13	1.038	1.164	19.00
2	57.48	62.52	8.65	10.58	10.58	1.016	1.196	17.62
2	68.03	73.01	14.18	8.09	8.09	1.007	1.220	16.37
1	69.71	70.25	12.39	8.78	8.76	1.009	1.214	16.69
2	78.57	82.72	23.86	5.71	5.68	1.002	1.239	15.28
1	78.95	79.51	19.89	6.46	6.48	1.003	1.233	15.63
1	87.75	88.18	33.68	4.31	4.30	1.001	1.248	14.70
1	93.86	94.02	52.71	2.80	2.80	1.000	1.257	14.11
1	96.95	96.98	69.93	2.01	2.04	1.000	1.261	13.82
1	100.00	100.00	100.00	1.25	1.26	1.000	1.265	-

WILSON PARAMETERS A = 0.200 B = 1.760 C = 1.000

Table 11. Acrylic Acid/Methyl Acrylate Vapor-Liquid Equilibrium Measurements by PTx Method at 80.0°C.

Run	Mole Percent AA			Pressure, kPa		Activity Coefficient		Relative Volatility
Half	Feed	Liquid	Vapor	Meas	Calc	AA	MA	MA/AA
3	0.00	0.00	0.00	101.06	101.06	1.859	1.000	11.51
3	7.16	7.22	0.70	95.13	94.80	1.530	1.007	11.04
3	14.02	14.12	1.48	89.83	89.44	1.349	1.022	10.98
3	23.52	23.66	2.76	82.38	82.20	1.203	1.049	10.91
3	33.49	33.64	4.53	74.32	74.36	1.117	1.080	10.69
2	41.18	41.24	6.30	68.51	68.06	1.077	1.104	10.44
3	44.76	44.90	7.33	64.70	64.91	1.062	1.116	10.29
2	49.57	49.66	8.90	60.68	60.70	1.047	1.130	10.09
2	61.45	61.57	14.37	49.77	49.65	1.021	1.165	9.54
1	67.32	67.47	18.30	44.07	43.93	1.014	1.181	9.26
2	74.70	74.82	25.02	36.95	36.59	1.007	1.200	8.91
1	79.80	79.93	31.49	31.37	31.38	1.004	1.212	8.66
1	89.19	89.27	50.26	21.45	21.65	1.001	1.233	8.23
1	94.77	94.81	69.58	15.71	15.78	1.000	1.244	7.98
1	100.00	100.00	100.00	10.22	10.22	1.000	1.254	-

WILSON PARAMETERS A = 0.288 B = 1.625 C = 1.000

Table 12. Acrylic Acid/Methyl
Measurements by PTxy

Pressure, kPa	Mole Pe Acrylic Liquid
91.84	11.23
75.34	33.66
58.54	52.23
41.35	71.32
26.28	85.13

Table 13. NMP/Chloroform Vapor-Liquid Equilibrium Measurements by PTx Method at 50.0°C.

Run	Mole Percent NMP			Pressure, kPa		Activity Coefficient		Relative Volatility
Half	Feed	Liquid	Vapor	Meas	Calc	NMP	$CHCl_3$	$CHCl_3$/NMP
2	0.00	0.00	0.00	69.93	69.93	0.073	1.000	2541.85
2	3.48	3.68	0.00	67.43	67.05	0.089	0.996	2085.91
2	6.14	6.76	0.00	64.65	64.27	0.104	0.987	1759.83
2	11.27	12.71	0.01	58.69	58.10	0.142	0.955	1255.59
2	21.21	24.06	0.05	44.88	44.51	0.245	0.844	646.52
2	27.97	31.32	0.11	35.69	35.67	0.334	0.750	422.18
2	37.91	41.68	0.31	23.93	24.33	0.488	0.603	233.45
1	50.06	50.62	0.70	16.15	16.52	0.633	0.483	144.56
1	59.97	60.37	1.66	10.27	10.29	0.782	0.372	90.41
1	69.81	70.15	3.72	6.38	6.11	0.898	0.287	60.90
1	80.03	80.17	8.25	3.67	3.39	0.970	0.229	44.94
1	89.93	89.98	19.27	1.72	1.67	0.997	0.197	37.62
1	95.04	95.06	34.72	0.98	0.98	1.000	0.190	36.20
1	97.77	97.78	54.93	0.64	0.64	1.000	0.189	36.06
1	100.00	100.00	100.00	0.36	0.36	1.000	0.190	-

REDLICH-KISTER PARAMETERS A = 0.073 B = 0.190 C = 0.230

D = 0.000

Table 14. NMP/Chloroform Vapor-Liquid Equilibrium Measurements by PTx Method at 100.0°C.

Run	Mole Percent NMP			Pressure, kPa		Activity Coefficient		Relative Volatility
Half	Feed	Liquid	Vapor	Meas	Calc	NMP	$CHCl_3$	$CHCl_3$/NMP
2	0.00	0.00	0.00	320.53	320.53	0.200	1.000	362.31
2	5.01	5.12	0.02	304.13	301.79	0.230	0.996	315.66
2	10.75	10.93	0.05	285.03	277.83	0.271	0.982	265.95
2	20.95	21.19	0.14	238.42	230.26	0.362	0.929	190.90
2	30.66	30.87	0.33	190.78	182.83	0.469	0.848	136.59
2	41.46	41.61	0.76	133.00	133.05	0.603	0.735	93.51
2	51.58	51.66	1.59	92.46	93.22	0.730	0.622	66.13
1	52.36	52.69	1.71	85.83	89.61	0.742	0.611	63.90
1	59.55	59.87	2.86	64.04	66.95	0.823	0.535	50.71
1	69.89	70.17	5.87	42.11	41.98	0.916	0.439	37.74
1	80.16	80.37	12.06	25.40	24.62	0.972	0.367	29.84
1	91.46	91.57	30.14	12.00	11.44	0.997	0.316	25.16
1	94.90	94.96	43.56	8.01	8.21	0.999	0.307	24.42
1	97.25	97.28	59.79	6.22	6.13	1.000	0.303	24.07
1	100.00	100.00	100.00	3.76	3.76	1.000	0.300	-

REDLICH-KISTER PARAMETERS A = 0.200 B = 0.300 C = 0.170

D = 0.000

Table 15. Infinite Dilution Activity Coefficients of Chloroform in NMP.

	50.0°C	100.0°C	150.0°C
γ^{∞}	0.19[(a)]	0.30[(a)]	0.43[(b)]

(a) Measured at 60 psia (414 kPa) with helium as the carrier gas.
(b) Measured at 200 psia (1379 kPa) with helium as the carrier gas.

Table 16. Pyridine/Furan Vapor-Liquid Equilibrium Measurements by PTx Method at 30.0°C.

Run Half	Mole Percent Pyridine Feed	Liquid	Vapor	Pressure, kPa Meas	Calc	Activity Coefficient Pyridine	Furan	Relative Volatility Furan/ Pyridine
2	0.00	0.00	0.00	97.90	97.90	1.047	1.000	26.21
2	4.85	4.88	0.19	92.35	93.20	1.041	1.000	26.37
2	9.85	9.90	0.41	87.58	88.38	1.036	1.001	26.54
2	14.74	14.81	0.65	83.76	83.70	1.032	1.001	26.69
2	19.69	19.77	0.91	78.74	79.00	1.027	1.002	26.85
2	24.63	24.73	1.20	74.21	74.31	1.024	1.003	27.00
2	34.54	34.65	1.91	65.19	64.98	1.017	1.006	27.30
1	44.09	44.15	2.79	56.27	56.08	1.012	1.009	27.57
2	44.49	44.60	2.84	56.37	55.66	1.012	1.009	27.58
1	49.06	49.14	3.37	51.65	51.41	1.010	1.011	27.71
1	54.04	54.14	4.07	47.09	46.73	1.008	1.013	27.85
2	54.47	54.58	4.13	47.02	46.32	1.008	1.013	27.86
1	58.98	59.10	4.91	42.45	42.08	1.006	1.015	27.98
1	64.00	64.13	5.98	37.58	37.37	1.005	1.018	28.11
1	69.03	69.16	7.36	33.01	32.64	1.003	1.020	28.24
1	74.08	74.20	9.20	28.12	27.90	1.002	1.023	28.37
1	79.18	79.30	11.85	23.28	23.09	1.001	1.026	28.50
1	84.29	84.37	15.87	18.35	18.29	1.001	1.029	28.62
1	89.48	89.54	22.95	13.27	13.38	1.000	1.032	28.75
1	94.82	94.85	38.94	8.25	8.33	1.000	1.035	28.87
1	100.00	100.00	100.00	3.41	3.41	1.000	1.038	-

NRTL PARAMETERS A = 0.128 B = -0.100 C = -1.000

Table 17. Pyridine/Furan Vapor-Liquid Equilibrium Measurements by PTx Method at 120.0°C.

Run	Mole Percent Pyridine			Pressure, kPa		Activity Coefficient		Relative Volatility
Half	Feed	Liquid	Vapor	Meas	Calc	Pyridine	Furan	Furan/ Pyridine
2	0.00	0.00	0.00	1136.25	1136.25	1.172	1.000	6.34
2	4.86	5.05	0.81	1074.19	1077.56	1.149	1.001	6.52
2	9.86	10.20	1.67	1016.28	1019.77	1.128	1.002	6.70
2	14.76	15.18	2.54	964.57	965.41	1.110	1.004	6.88
2	19.71	20.17	3.46	911.48	912.22	1.094	1.008	7.04
2	24.67	25.13	4.45	863.22	860.40	1.079	1.011	7.21
2	34.59	34.99	6.67	763.24	759.81	1.056	1.021	7.53
1	44.13	44.22	9.21	670.37	667.40	1.039	1.032	7.82
2	44.56	44.81	9.39	664.92	661.59	1.038	1.033	7.83
1	49.11	49.29	10.87	620.87	617.23	1.031	1.039	7.97
1	54.10	54.37	12.79	568.40	567.12	1.024	1.046	8.12
2	54.56	54.63	12.90	569.50	564.54	1.024	1.047	8.13
1	59.05	59.39	15.03	518.90	517.61	1.019	1.054	8.27
1	64.07	64.47	17.73	470.29	467.72	1.014	1.062	8.42
1	69.10	69.55	21.06	419.40	417.79	1.010	1.070	8.56
1	74.14	74.60	25.23	369.07	368.03	1.007	1.079	8.70
1	79.23	79.68	30.72	319.57	317.92	1.004	1.088	8.85
1	84.32	84.73	38.18	267.38	268.03	1.002	1.097	8.98
1	89.50	89.81	49.16	215.94	217.52	1.001	1.107	9.12
1	94.83	95.01	67.28	164.30	165.76	1.000	1.117	9.26
1	100.00	100.00	100.00	115.83	115.83	1.000	1.127	-

NRTL PARAMETERS A = 0.210 B = -0.100 C = -1.000

Table 18. NMP/p-Xylene Vapor-Liquid Equilibrium Measurements by PTx Method at 100°C.

Run	Mole Percent NMP			Pressure, kPa		Activity Coefficient		Relative Volatility
Half	Feed	Liquid	Vapor	Meas	Calc	NMP	p-Xylene	p-Xylene/ NMP
2	0.00	0.00	0.00	31.45	31.74	2.663	1.000	3.47
2	2.63	2.63	0.70	31.38	31.15	2.416	1.001	3.83
2	5.03	5.04	1.26	31.03	30.64	2.231	1.004	4.16
2	10.28	10.30	2.30	29.81	29.60	1.923	1.017	4.89
2	20.93	20.97	4.00	27.53	27.61	1.538	1.059	6.36
2	31.78	31.82	5.62	25.45	25.56	1.319	1.118	7.84
2	46.06	46.09	8.10	22.29	22.50	1.158	1.214	9.70
1	46.01	46.03	8.09	22.33	22.51	1.158	1.214	9.69
2	56.20	56.23	10.49	19.81	19.93	1.091	1.292	10.96
1	56.72	56.76	10.64	19.50	19.79	1.088	1.296	11.02
1	70.66	70.70	16.01	15.34	15.54	1.035	1.415	12.66
1	80.71	80.76	23.35	11.97	11.90	1.013	1.508	13.78
1	90.20	90.24	38.47	8.12	7.97	1.003	1.601	14.78
1	94.73	94.76	54.24	5.97	5.91	1.001	1.647	15.25
1	97.41	97.42	70.86	4.82	4.65	1.000	1.674	15.52
1	100.00	100.00	100.00	3.38	3.38	1.000	1.702	-

WILSON PARAMETERS: A = 0.320 B = 1.160 C = 1.000

Table 19. NMP/p-Xylene Vapor-Liquid Equilibrium Measurements by PTx Method at 200°C.

Run Half	Mole Percent NMP Feed	Mole Percent NMP Liquid	Mole Percent NMP Vapor	Pressure, kPa Meas	Pressure, kPa Calc	Activity Coefficient NMP	Activity Coefficient p-Xylene	Relative Volatility p-Xylene/NMP
0	0.00	0.00	0.00	414.60	414.60	2.069	1.000	1.90
2	0.00	0.00	0.00	414.60	414.59	2.070	1.000	1.90
2	2.63	2.65	1.33	409.96	408.66	1.944	1.001	2.03
2	5.04	5.09	2.44	404.37	403.21	1.844	1.003	2.14
2	10.28	10.38	4.61	392.14	391.37	1.666	1.011	2.40
2	20.94	21.10	8.43	365.93	367.18	1.418	1.042	2.91
2	31.79	31.98	12.09	339.94	341.48	1.260	1.087	3.42
2	46.07	46.22	17.44	303.75	304.08	1.135	1.162	4.07
2	46.52	46.57	17.59	303.06	303.07	1.132	1.164	4.08
2	55.79	55.87	21.97	275.92	275.27	1.081	1.221	4.50
2	56.75	56.90	22.52	270.11	272.00	1.077	1.228	4.54
2	70.68	70.94	32.18	223.42	223.27	1.030	1.327	5.15
2	80.73	81.01	43.40	184.51	183.08	1.012	1.404	5.56
2	90.21	90.41	61.33	141.82	141.25	1.003	1.482	5.94
2	94.74	94.87	75.13	120.48	119.88	1.001	1.521	6.12
2	97.41	97.48	86.13	106.33	106.91	1.000	1.544	6.22
2	100.00	100.00	100.00	94.04	94.04	1.000	1.567	-

WILSON PARAMETERS: A = 0.420 B = 1.140 C = 1.000

Table 20. DMF/1,3-Butadiene Vapor-Liquid Equilibrium Measurements by PTx Method at 40.0°C.

Run Half	Mole Percent DMF Feed	Mole Percent DMF Liquid	Mole Percent DMF Vapor	Pressure, kPa Meas	Pressure, kPa Calc	Activity Coefficient DMF	Activity Coefficient Butadiene	Relative Volatility Butadiene/DMF
2	0.00	0.00	0.00	437.05	437.05	5.589	1.000	47.05
2	2.57	2.62	0.05	425.89	425.72	4.544	1.003	58.26
2	5.25	5.36	0.08	416.30	415.99	3.792	1.010	70.58
2	10.16	10.35	0.12	402.10	401.62	2.913	1.033	94.45
2	20.42	20.73	0.18	378.31	377.90	2.001	1.105	148.30
2	31.19	31.54	0.22	353.28	354.09	1.560	1.205	209.35
1	48.59	48.66	0.30	306.75	306.98	1.231	1.409	315.25
2	46.66	46.92	0.29	310.81	312.58	1.253	1.385	303.96
2	59.14	59.29	0.37	266.07	267.61	1.126	1.562	387.28
1	60.31	60.54	0.39	262.48	262.32	1.117	1.582	396.13
1	73.07	73.44	0.56	199.33	198.55	1.046	1.803	492.32
1	80.01	80.39	0.74	157.34	156.29	1.024	1.938	548.48
1	90.19	90.59	1.49	83.06	82.72	1.005	2.160	637.45
1	94.96	95.20	2.83	44.15	44.57	1.001	2.270	680.51
1	100.00	100.00	100.00	1.29	1.29	1.000	2.391	-

WILSON PARAMETERS: A = 0.190 B = 0.940 C = 1.000

Table 21. DMF/1,3-Butadiene Vapor-Liquid Equilibrium Measurements by PTx Method at 78.82°C.

Run Half	Mole Percent DMF Feed	Liquid	Vapor	Pressure, kPa Meas	Calc	Activity Coefficient DMF	Butadiene	Relative Volatility Butadiene/DMF
2	0.00	0.00	0.00	1125.21	1125.21	4.980	1.000	17.39
2	2.56	2.70	0.13	1095.57	1091.45	4.158	1.002	21.11
2	5.25	5.52	0.23	1064.54	1061.97	3.538	1.009	25.23
2	10.16	10.62	0.36	1021.11	1018.10	2.785	1.030	33.18
2	20.42	21.12	0.52	948.71	946.40	1.966	1.099	51.28
2	31.19	31.95	0.65	877.70	877.78	1.551	1.196	72.21
1	46.66	47.18	0.84	761.17	766.02	1.255	1.371	105.78
2	48.60	48.67	0.86	750.49	753.48	1.236	1.391	109.34
2	59.14	59.40	1.06	647.76	650.91	1.129	1.546	136.91
1	60.31	60.75	1.09	641.28	636.35	1.119	1.567	140.64
1	73.07	73.82	1.54	478.01	474.16	1.047	1.796	179.91
1	80.00	80.78	2.02	373.21	369.90	1.024	1.936	203.66
1	90.19	90.76	3.90	195.19	196.26	1.005	2.162	241.81
1	94.96	95.31	7.22	105.49	106.97	1.001	2.277	261.00
1	100.00	100.00	100.00	7.77	7.77	1.000	2.402	-

WILSON PARAMETERS: A = 0.220 B = 0.908 C = 1.000

Table 22. Diethanolamine/Ethylene Glycol Vapor-Liquid Equilibrium Measurements by PTx Method at 120.0°C.

Run Half	Mole Percent DEA Feed	Liquid	Vapor	Pressure, kPa Meas	Calc	Activity Coefficient DEA	EG	Relative Volatility EG/DEA
2	0.00	0.00	0.00	5.37	5.37	0.686	1.000	40.31
2	2.66	2.66	0.07	5.24	5.23	0.700	1.000	39.50
2	4.97	4.97	0.13	5.10	5.11	0.711	0.999	38.83
2	9.92	9.92	0.29	4.84	4.83	0.737	0.996	37.40
2	19.82	19.82	0.71	4.17	4.27	0.785	0.985	34.72
2	29.86	29.86	1.30	3.63	3.69	0.831	0.967	32.21
1	45.85	45.85	2.88	2.77	2.77	0.895	0.924	28.58
1	55.80	55.80	4.54	2.24	2.21	0.929	0.889	26.54
1	70.71	70.71	9.23	1.46	1.43	0.968	0.828	23.73
1	80.71	80.72	15.98	0.99	0.96	0.986	0.782	22.01
1	90.40	90.40	31.52	0.54	0.55	0.996	0.735	20.46
1	95.47	95.47	51.71	0.32	0.36	0.999	0.709	19.69
1	97.54	97.54	67.13	0.26	0.28	1.000	0.698	19.38
1	100.00	100.00	100.00	0.19	0.19	1.000	0.685	-

WILSON PARAMETERS A = 1.200 B = 1.195 C = 1.000

Table 23. Diethanolamine/Ethylene Glycol Vapor-Liquid Equilibrium Measurements by PTx Method at 160.0°C.

Run Half	Mole Percent DEA Feed	Liquid	Vapor	Pressure, kPa Meas	Calc	Activity Coefficient DEA	EG	Relative Volatility EG/DEA
2	0.00	0.00	0.00	29.09	29.09	0.838	1.000	15.59
2	2.80	2.81	0.19	28.40	28.32	0.847	1.000	15.45
2	5.68	5.69	0.39	27.63	27.52	0.855	0.999	15.30
2	11.35	11.38	0.85	25.95	25.93	0.871	0.998	15.01
2	19.64	19.67	1.65	23.59	23.58	0.893	0.993	14.60
2	30.03	30.07	2.96	20.62	20.60	0.918	0.984	14.10
1	45.12	45.14	5.79	16.08	16.30	0.948	0.965	13.40
2	45.21	45.24	5.81	16.34	16.28	0.949	0.965	13.40
1	54.95	54.98	8.61	13.36	13.55	0.965	0.948	12.97
2	54.98	54.99	8.61	13.71	13.54	0.965	0.948	12.97
1	69.99	70.03	15.93	9.54	9.48	0.984	0.917	12.33
1	79.50	79.53	24.55	6.90	7.03	0.993	0.895	11.94
1	89.72	89.74	43.12	4.54	4.53	0.998	0.868	11.53
1	95.15	95.16	63.47	3.23	3.27	1.000	0.853	11.32
1	97.09	97.10	74.85	2.87	2.82	1.000	0.847	11.25
1	100.00	100.00	100.00	2.18	2.18	1.000	0.838	-

WILSON PARAMETERS A = 1.090 B = 1.090 C = 1.000

Table 24. Comparison of Pure Component Vapor Pressures to Literature Values.

System	Compound	Temp (°C)	Vapor Pressure (kPa) Measured	Literature	Source
1	Benzoic Acid	150	3.18	2.95	a
		200	22.98	22.30	a
	Benzonitrile	150	33.67	32.21	b
		200	126.52	126.55	b
2	Methyl tert-butyl Ether	30	40.20	40.31	c
		90	282.54	283.30	c
	Propylene Oxide	30	86.87	86.72	a
		90	544.61	536.34	d
3	Acetic Anhydride	40	1.53	1.77	a
		140	107.14	105.4	a
	Methyl Iodide	40	94.80	93.17	a
		140	1161.07	1173.3	a
4	DMF	30	0.69	0.73	a
		60	3.59	3.89	a
	Water	30	4.23	4.22	a
		60	19.99	19.92	a
5	Acrylic Acid	40	1.25	1.35	a
		80	10.22	10.90	a
	Methyl Acrylate	40	22.37	-	
		80	101.06	-	

Table 24. Comparison of Pure Component Vapor Pressures to Literature Values (continued).

System	Compound	Temp (°C)	Vapor Pressure (kPa) Measured	Vapor Pressure (kPa) Literature	Source
6	NMP	50	0.36	-	
		100	3.76	-	
	Chloroform	50	69.93	69.42	b
		100	320.53	314.12	b
7	Pyridine	30	3.41	3.62	a
		120	115.83	115.97	a
	Furan	30	97.90	96.43	a
		120	1136.25	1120.71	a
8	NMP	100	3.38	-	
		200	94.04	-	
	p-Xylene	100	31.45	32.06	a
		200	414.60	412.90	a
9	DMF	40	1.29	1.34	d
		78.82	7.77	9.19	d
	1,3-Butadiene	40	437.05	434.33	a
		78.82	1125.21	1110.75	a
10	Diethanolamine	120	0.19	0.226	d
		160	2.18	2.00	d
	Ethylene Glycol	120	5.37	5.65	a
		160	29.09	28.84	a

(a) Antoine equation from Reference 2.
(b) Harlacher equation from Reference 2.
(c) From Antoine equation developed from Antoine equation for Methyl Butyl Ether (Reference 2).
(d) Antoine equation from Reference 4.

APPENDIX

Activity Coefficient Equations

MODIFIED WILSON

$$\ln \gamma_1 = -\ln (x_1 + Ax_2) + x_2 \left(\frac{A}{x_1 + Ax_2} - \frac{B}{Bx_1 + x_2} \right)$$

$$\ln \gamma_2 = -\ln (x_1 + Bx_2) - x_2 \left(\frac{A}{x_1 + Ax_2} - \frac{B}{Bx_1 + x_2} \right)$$

$$\gamma_1 = \gamma_1^C$$

$$\gamma_2 = \gamma_2^C$$

Note: When C=1 this equation becomes the standard Wilson Equation.

NRTL

$$G_{12} = e^{-C \cdot A}$$

$$G_{21} = e^{-C \cdot B}$$

$$\ln \gamma_1 = x_2^2 \left[B \left(\frac{G_{21}}{x_1 + x_2 G_{21}} \right)^2 + \frac{AG_{12}}{(x_2 + x_1 G_{12})^2} \right]$$

$$\ln \gamma_2 = x_1^2 \left[A \left(\frac{G_{12}}{x_2 + x_1 G_{12}} \right)^2 + \frac{BG_{21}}{(x_1 + x_2 G_{21})^2} \right]$$

REDLICH - KISTER (3 Parameters)

$$RT \ln\gamma_1 = a_1 x_2^2 + b_1 x_2^3 + c_1 x_2^4$$

$$RT \ln\gamma_2 = a_2 x_1^2 + b_2 x_1^3 + c_2 x_1^4$$

$a_1 = A + 3B + 5C$ $\quad$ $a_2 = A - 3B + 5C$

$b_1 = -4(B + 4C)$ $\quad$ $b_2 = 4(B-4C)$

$c_1 = 12C$ $\quad$ $c_2 = 12C$

x_i = liquid mole fraction of component i.

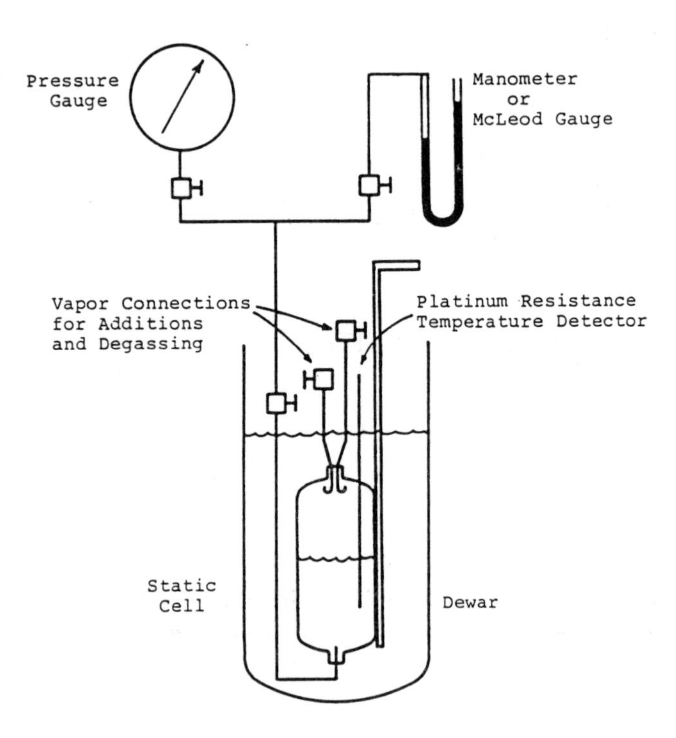

Figure 1. Schematic of agitated static cell used for PTx measurements.

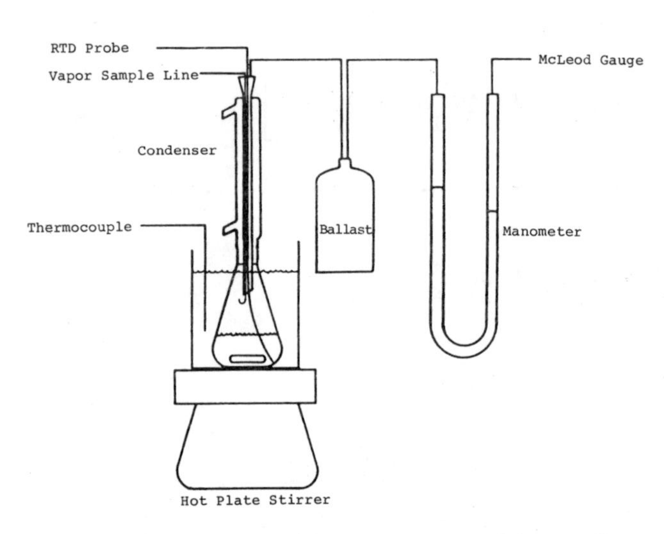

Figure 2. Glass still used for low pressure PTx and PTxy measurements.

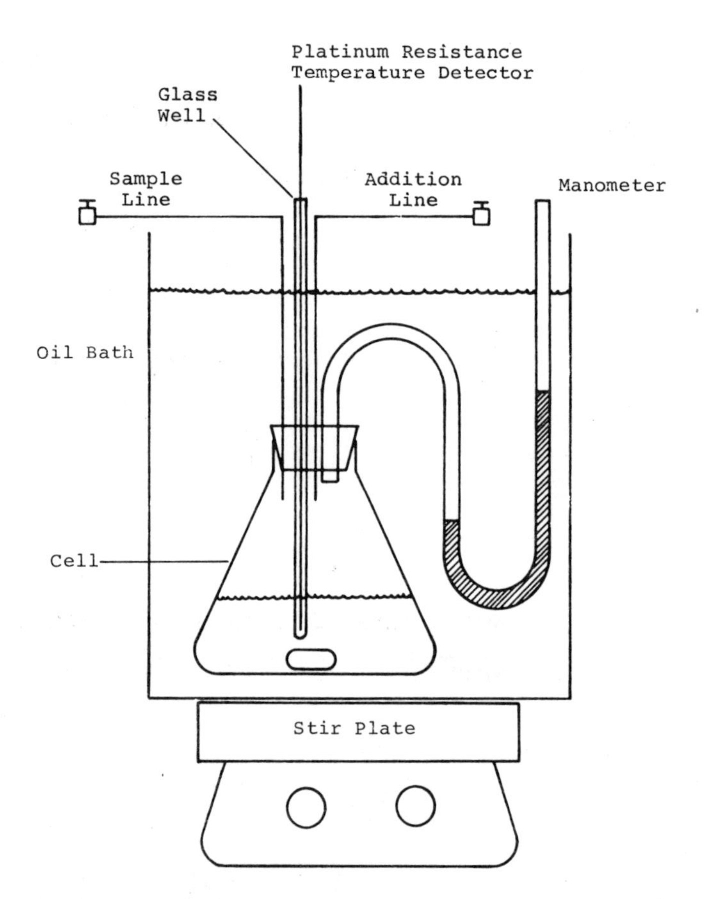

Figure 3. Schematic of apparatus used for PTx and PTxy measurements.

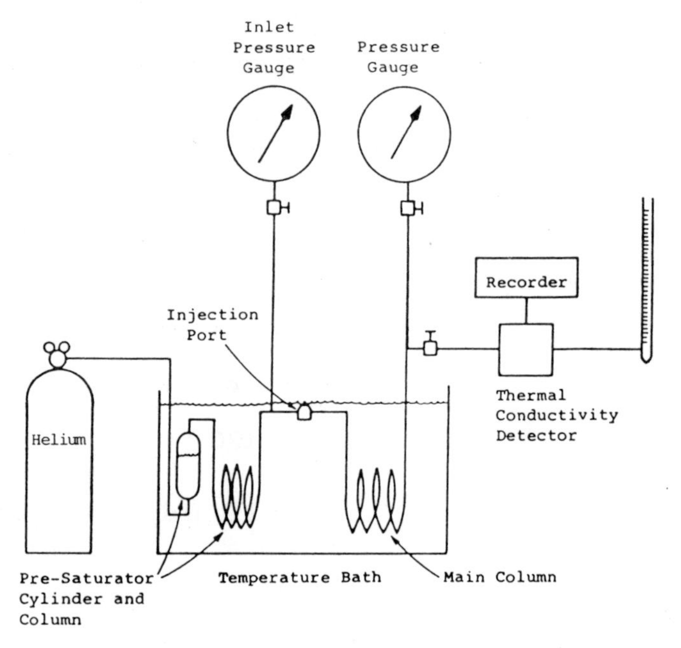

Figure 4. Schematic of gas chromatographic apparatus used for measuring infinite dilution activity coefficients.

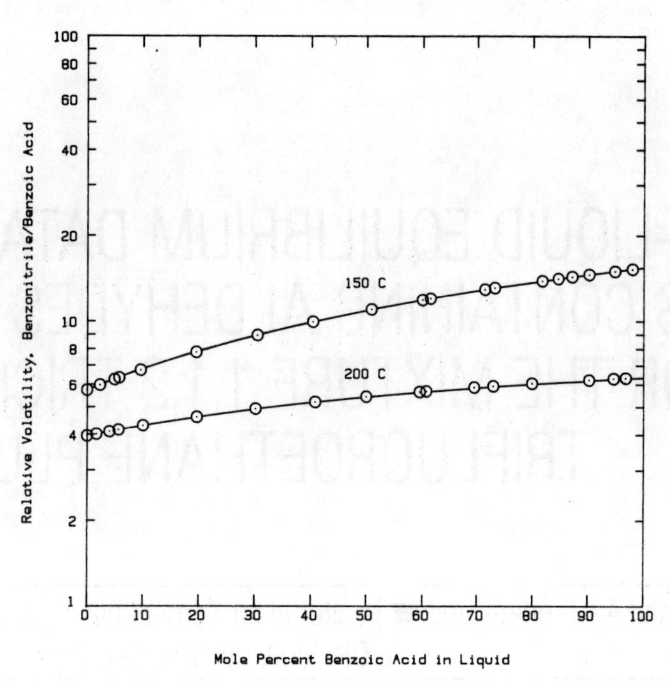

Figure 5. Relative volatility of benzonitrile/benzoic acid from PTx data.

VAPOR-LIQUID EQUILIBRIUM DATA FOR BINARY MIXTURES CONTAINING ALDEHYDES AND ESTERS AND FOR THE MIXTURE 1,1,2 TRICHOLORO-1,2,2 TRIFLUOROETHANE PLUS N-HEXANE

G. Radnai*, P. Rasmussen and Aa. Fredenslund ■ Instituttet for Kemiteknik, Technical University of Denmark, DK-2800, Lyngby, Denmark

Vapor-liquid equilibrium (VLE) data have been measured for binary mixtures 1,1,2 trichloro-1,2,2 trifluoro-ethane (3C&3F-ethane) plus n-hexane. A total of five systems have been investigated in the temperature range of 303 to 343 K and from atmospheric pressure and below. The measurements were carried out in a still similar to that of Dvorak and Boublik. The data were reduced using Barker's method, and they were found thermodynamically consistent. In addition, UNIFAC parameters were determined for the aldehyde/ester functional group interactions, and the system 3C&3F-ethane plus n-hexane was correlated using the UNIQUAC equation.

INTRODUCTION

The UNIFAC group-contribution model for the prediction of activity coefficients [1-3] is dependent on reliable vapor-liquid equilibrium data. Until now parameters describing the energetic interactions between the aldehyde and the ester groups have not been available due to insufficient, reliable data. In this work, VLE data are reported for four different aldehyde plus ester systems, and ester-aldehyde UNIFAC group interaction parameters are determined. In addition, VLE data are reported for the system 1,1,2 trichloro - 1,2,2 trifluoro-ethane (3Cℓ3-F-ethane) plus n-hexane, and the data are correlated with the UNIQUAC model.

The following isothermal data are reported:

3Cℓ3F-ethane - n-hexane 30 and 40 °C
Propionaldehyde-methyl acetate 30 and 40 °C
Propionaldehyde-ethyl acetate 30 and 40 °C
Methyl acetate-butyraldehyde 40 and 50 °C
Butyraldehyde-propyl-acetate 50 and 60 °C

*on leave from Technical University of Budapest, Hungary.

EXPERIMENTAL TECHNIQUE

Apparatus. The VLE data were measured in a glass Dvorak-Boublik still [4], which provides for circulation of both the vapor and liquid phases. The equilibrium temperature was measured with a Pt resistance thermometer via a Müller bridge, System Teknik AB, S1118 with an accuracy of ± 0.01 K. The pressure was measured to within 0.1 mmHg with a mercury-filled U-tube manometer (inner diameter 20mm) using a kathetometer.

Materials. Table I shows the chemicals used and the steps taken for further purification. After purification, the purity of th chemicals was, based on gas chromatographic analyses, in all cases greater than 99.9%.

Analysis. A Hewlett-Packard 5840A gas chromatograph (GC) was used for the analyses. Table II summarizes the operating conditions which were found to give the best separation for each system.

Pure-Component Vapor Pressures. The Dvorak-Boublik still was used to determinde the vapor pressures of the pure components. This was done in order to check the purity of the components and to ascertain the optimal operating conditions the the VLE measurements. Table III summarizes the fit of the experimental data to the Antoine equation:

$$\log P_i^\circ \text{ (mmHg)} = A_i - B_i/(t(^\circ C)+C_i)$$

VLE Measurements. The systems studied are those shown in the introduction. A total of 10 isothermal PTxy data sets were determined. The experimental results are shown in Table IV.

Data reduction.

The equilibrium equation used for the data reduction is

$$PY_i\Phi_i = x_i\gamma_i P_i^\circ\Phi_i^\circ \quad i = 1,2$$

The experimental data were checked for thermodynamic consistency using Barker's method as described in detail by Fredenslund et al. [1]. The fugacity coefficients Φ_i are computed from the virial equation of state truncated after the second term, and with second virial coefficients predicted from the method of Hayden and O'Connell [6]. Table V summarizes the results.

In addition, UNIFAC parameters were obtained for the aldehyde-ester interaction using the maximum likelihood procedure described i [5]. The results are shown in Table VI, which also gives the new group-interaction parameters. The group R_k-and Q_k-values (volume and surface area constants) are those given in references [2,3].

UNIQUAC parameters for the system 3Cl3F-ethane plus n-hexane are given in Table VII.

CONCLUSION

The results from Barker's consistency test show that the mean deviations between the experimental and calculated values of P and y are comparable to the experimental uncertainties. All experimental data sets may therefore be considered to be thermodynamically consistent and the data sets have been included in the data base for obtaining UNIFAC and UNIQUAC parameters.

ACKNOWLEDGEMENT

The authors are grateful to DIPPR of the American Institute of Chemical Engineers for financial support of the project and to Ole Persson for assistance with the laboratory work.

GLOSSARY

a_{12}, a_{21}	UNIFAC parameters in Kelvin
A, B, C	constants in the Antoine equation
P	pressure in mm Hg
P_i°	vapor pressure in mm Hg of component i
q_i	Surface area parameter of component i
r_i	volume parameter of component i
t	temperature °C
T	temperature in Kelvin
x_i	liquid-phase mole fraction of component i
y_i	vapor-phase mole fraction of component i
Φ_i	fugacity coefficient of component i in mixture
Φ_i°	fugacity coefficient of pure component i
γ_i	activity coefficient

LITTERATURE CITED

[1] Fredenslund, Aa., Gmehling, J., Rasmussen, P.
"Vapor-Liquid Equilibria using UNIFAC"; Elsevier:
Amsterdam, 1977

[2] Gmehling, J., Rasmussen, P., Fredenslund, Aa.
Ind.Eng.Chem.Process Des.Dev. 1982, 21, 118

[3] Macedo, E.A., Weidlich, U., Gmehling, J., Rasmussen, P.
Ind.Eng.Chem.Process Des.Dev. 1983, 22, 676

[4] Dvorak, K., Boublik, T. Collect.-Czech. Chem. Commun.
1963, 28, 1249

[5] Kemeny, S., Skjold-Jørgensen, S.,Manczinger, J., Toth,
K. AIChE J. 1982, 28, 20

[6] Hayden, J.G., O'Connell, J.P. Ind.Eng.Chem. Process.
Des.Dev. 1975, 14, 209.

TABLE I. Pure Components

Materials	Firm	Purity (as measured by G.C.)
1) n-Hexane	Merck	>99%
2) 3Cl3F-ethane	Mallincrodt	99,5%
3) Propionaldehyde	Merck	98%
4) Butyraldehyde	Merck	98%
5) Methyl acetate	Merck	>99%
6) Ethyl acetate	Merck	>99%
7) Propyl acetate	Fluka	>99%

Further purification:

Materials	Methods
Propionaldehyde	drying ($MgSO_4$), destillation, molecular sieve type 3A.
Butyraldehyde	- " -
Propyl acetate	distillation

TABLE II

Gas Chromatograph Operating Conditions

Chromatograph	: Hewlett-Packard 5840 A
Detector	: Thermal conductivity
Carrier Gas	: Helium, 25 ml/min
Slope sensitivity	: 10.0
Attenuation	: 2 q

	1	2	3	4	5
Column: S.S.,D=1/8"					
length [m]:	1,4	2	1,4	1,4	1,4
Packing	PORAPAQ Q 50/80 mesh				
Inject. temp(°C)	200	200	200	200	200
Detect.temp.(°C)	250	250	250	250	250
Oven temp. (°C)	210	210	210	210	230
Sample volume (μl)	0,3	0,4	0,4	0,4	0,5

System		Retention time (min)	
I	II	I	II
1) Trichlorotrifluoroethane-	n.Hexane	1.41	2.12
2) Propionaldehyde	- Methyl acetate	3.16	3.83
3) Propionaldehyde	- Ethyl acetate	1.16	1.96
4) Methyl acetate	- Butyraldehyde	1.37	1.92
5) Butyraldehyde	- Propyl acetate	1.30	2.13

TABLE III

$$(\text{STD.DEV} = \left[(\text{PEXP}-\text{PCAL})^2/(\text{NDP}-3\right]^{0,5})$$

Pure-Component Vapor Pressure Measurements

Component	No. of data points	STD.DEV	New Antoine Parameters		
1) n-Hexane	15	0.18	6.942959	1197.174	226.116
2) 3Cℓ3F-ethane	12	0.13	6.679522	1003.150	216.365
3) Propionaldehyde	13	0.17	6.642106	929.446	199.030
4) Butyraldehyde	14	0.24	6.951817	1196.996	219.202
5) Methyl acetate	12	0.30	7.411301	1342.617	239.471
6) Ethyl acetate	14	0.27	7.374986	1400.306	234.505
7) Propyl acetate	13	0.36	6.980542	1260.975	206.231

TABLE IV.

Experimental VLE Data (TEXP is in K, PEXP in mm Hg)

3Cℓ3F-ethane (1) - n-Hexane (2) at 303. K

X1EXP	TEXP	Y1EXP	PEXP
0.0000	303.15	0.0000	187.10
0.0616	303.18	0.1486	210.76
0.1288	303.13	0.2683	229.76
0.1667	303.16	0.3354	242.45
0.2306	303.11	0.4177	260.35
0.3233	303.20	0.5306	285.89
0.3865	303.15	0.5899	301.39
0.4504	303.15	0.6468	313.57
0.5222	303.17	0.6946	329.06
0.5814	303.13	0.7341	338.67
0.6311	303.15	0.7675	349.36
0.6856	303.21	0.8002	358.65
0.7342	303.11	0.8325	365.86
0.7827	303.15	0.8581	375.45
0.8578	303.18	0.9074	386.24
0.9307	303.15	0.9501	395.84
1.0000	303.15	1.0000	405.65

TABLE IV, cont.

3Cℓ3F-ethane (1) - n-Hexane (2) at 313. K

X1EXP	TEXP	Y1EXP	PEXP
0.0000	313.15	0.0000	279.44
0.0555	313.12	0.1460	310.48
0.1223	313.18	0.2527	338.59
0.1680	313.15	0.3334	356.53
0.2334	313.15	0.4191	380.90
0.3173	313.15	0.5197	415.50
0.3807	313.12	0.5798	435.62
0.4531	313.18	0.6399	457.84
0.5153	313.15	0.6893	475.94
0.5791	313.10	0.7300	490.84
0.6298	313.12	0.7641	504.18
0.6820	313.15	0.7969	516.66
0.7328	313.17	0.8273	530.08
0.7806	313.11	0.8533	539.14
0.8572	313.15	0.9054	557.60
0.9317	313.15	0.9503	571.36
1.0000	313.15	1.0000	583.54

TABLE IV, cont.

Methyl acetate - Butyraldehyde at 313. K

X1EXP	TEXP	Y1EXP	PEXP
0.0000	313.15	0.0000	215.64
0.0753	313.10	0.1419	230.22
0.1080	313.17	0.2080	236.10
0.1569	313.11	0.2669	245.02
0.2326	313.15	0.3762	261.17
0.2987	313.16	0.4548	275.13
0.3680	313.09	0.5264	285.80
0.4346	313.17	0.5966	300.11
0.5083	313.15	0.6607	313.98
0.5627	313.11	0.7049	322.28
0.6137	313.15	0.7489	333.12
0.6474	313.17	0.7782	340.97
0.6974	313.17	0.8113	350.01
0.7393	313.15	0.8364	358.68
0.8400	313.18	0.9048	378.11
0.9246	313.15	0.9585	393.65
1.0000	313.15	1.0000	405.62

TABLE IV, cont.

Methyl acetate - Butyraldehyde at 323. K

X1EXP	TEXP	Y1EXP	PEXP
0.0000	323.15	0.0000	320.48
0.0759	323.15	0.1481	340.20
0.1052	323.19	0.1863	350.02
0.1533	323.16	0.2606	365.68
0.2276	323.18	0.3640	385.70
0.2975	323.08	0.4408	403.85
0.3526	323.10	0.5067	419.20
0.4278	323.10	0.5832	441.77
0.5092	323.15	0.6609	462.10
0.5559	323.21	0.6968	477.06
0.6187	323.15	0.7506	492.74
0.6457	323.12	0.7702	501.48
0.6965	323.16	0.8104	512.03
0.7322	323.15	0.8355	521.35
0.8435	323.15	0.9110	551.01
0.9300	323.15	0.9606	575.12
1.0000	323.15	1.0000	593.43

TABLE IV, cont.

Propionaldehyde - Ethyl acetate at 303. K

X1EXP	TEXP	Y1EXP	PEXP
0.0000	303.15	0.0000	119.96
0.0124	303.15	0.0404	122.90
0.0338	303.15	0.0946	127.35
0.0850	303.15	0.2240	141.10
0.1415	303.11	0.3316	153.91
0.1873	303.18	0.4165	168.44
0.2265	303.15	0.4903	177.14
0.2755	303.17	0.5510	193.64
0.3077	303.21	0.5962	213.14
0.4283	303.17	0.7038	238.20
0.5009	303.17	0.7611	255.70
0.5893	303.19	0.8220	279.41
0.6612	303.20	0.8585	298.36
0.7457	303.13	0.8996	320.02
0.8128	303.17	0.9289	334.97
0.8732	303.10	0.9520	350.40
0.9395	303.21	0.9819	369.50
1.0000	303.15	1.0000	389.60

TABLE IV, cont.

Propionaldehyde - Ethyl acetate at 313. K

X1EXP	TEXP	Y1EXP	PEXP
0.0000	313.15	0.0000	187.85
0.0113	313.18	0.0362	191.63
0.0293	313.13	0.0780	197.97
0.0841	313.15	0.2110	219.20
0.1252	313.10	0.2929	233.37
0.1844	313.19	0.3899	255.56
0.2250	313.15	0.4670	272.08
0.2599	313.15	0.5150	287.19
0.2994	313.18	0.5626	303.05
0.4164	313.20	0.6818	350.34
0.5035	313.21	0.7510	384.36
0.5752	313.10	0.8021	409.12
0.6600	313.12	0.8518	441.10
0.7393	313.17	0.8903	472.33
0.8108	313.15	0.9193	495.77
0.8716	313.12	0.9500	519.30
0.9355	313.17	0.9780	542.44
1.0000	313.15	1.0000	570.64

TABLE IV, cont.

Propionaldehyde - Methyl acetate at 303. K

X1EXP	TEXP	Y1EXP	PEXP
0.0000	303.15	0.0000	268.93
0.0752	303.11	0.0999	274.83
0.1354	303.16	0.1813	282.19
0.1960	303.21	0.2525	290.69
0.2727	303.19	0.3360	299.85
0.3237	303.15	0.3973	305.83
0.3931	303.11	0.4701	316.17
0.4707	303.17	0.5512	324.10
0.5528	303.17	0.6274	334.04
0.6495	303.18	0.7195	346.90
0.7422	303.20	0.7979	357.27
0.8158	303.14	0.8537	365.43
0.8741	303.17	0.9030	372.57
0.9250	303.15	0.9470	377.96
1.0000	303.15	1.0000	389.59

TABLE IV, cont.

Propionaldehyde - Methyl acetate at 313. K

X1EXP	TEXP	Y1EXP	PEXP
0.0000	313.15	0.0000	405.62
0.0766	313.09	0.1079	415.70
0.1315	313.09	0.1759	426.08
0.1913	313.15	0.2515	436.41
0.2648	313.16	0.3385	448.71
0.3189	313.10	0.3963	456.99
0.3382	313.17	0.4618	470.76
0.4657	313.11	0.5459	483.29
0.5458	313.17	0.6184	497.64
0.6423	313.09	0.7037	513.36
0.7420	313.18	0.7893	528.10
0.8140	313.09	0.8495	538.28
0.8719	313.10	0.8963	547.70
0.9242	313.17	0.9361	556.94
1.0000	313.15	1.0000	570.64

TABLE IV, cont.

Butyraldehyde - Propyl acetate at 323. K

X1EXP	TEXP	Y1EXP	PEXP
0.0000	323.15	0.0000	114.18
0.0597	323.18	0.1600	126.39
0.1220	323.17	0.2719	140.96
0.1748	323.16	0.3572	150.38
0.2534	323.15	0.4808	165.33
0.3194	323.16	0.5524	177.61
0.4012	323.17	0.6399	193.94
0.4964	323.16	0.7256	215.68
0.5709	323.15	0.7797	229.56
0.6538	323.15	0.8339	249.73
0.7589	323.10	0.9001	269.56
0.8431	323.13	0.9342	286.56
0.9200	323.18	0.9676	304.20
1.0000	323.15	1.0000	320.48

TABLE IV, cont.

Butyraldehyde - Propyl acetate at 333. K

X1EXP	TEXP	Y1EXP	PEXP
0.0000	333.15	0.0000	174.66
0.0671	333.19	0.1688	193.24
0.1200	333.20	0.2653	208.05
0.1804	333.13	0.3579	224.69
0.2583	333.15	0.4677	246.51
0.3288	333.13	0.5541	267.17
0.3945	333.14	0.6238	286.52
0.4864	333.12	0.7151	313.34
0.5645	333.10	0.7750	334.12

0.6643	333.16	0.8376	362.97
0.7531	333.11	0.8889	387.22
0.8493	333.15	0.9387	418.61
0.9302	333.17	0.9689	439.99
1.0000	333.15	1.0000	461.65

Table V

Average deviations resulting from consistency tests

(Barker's method)

System	T,K	$\lvert\Delta P\rvert$ mmHg	$\lvert\Delta y\rvert$
3Cℓ3F-ethane - n-hexane	303 313	0.73 1.16	0.0064 0.0040
Propionaldehyde - Methyl acetate	303 313	0.59 0.64	0.0087 0.0058
Propionaldehyde - Ethyl acetate	303 313	1.20 0.97	0.0052 0.0035
Methyl acetate - Butyraldehyde	313 323	0.84 0.95	0.0065 0.0070
Butyraldehyde - Propyl acetate	323 333	0.82 0.62	0.0043 0.0053

Table VI

New UNIFAC group-interaction parameters

and results from the correlation

of the experimental data.

"CCOO-CHO"

System	T,K	No. of data points	Average Deviations $\lvert\Delta x\rvert$	$\lvert\Delta T\rvert$ K	$\lvert\Delta y\rvert$	$\lvert\Delta P\rvert$ mmHg
Propionaldehyde-Methyl acetate	303-313	30	0.0048	0.06	0.0043	0.44
Propionaldehyde-Ethyl acetate	303-313	36	0.0084	0.05	0.0063	0.14
Methyl acetate-Butyraldehyde	313-323	34	0.0029	0.04	0.0058	0.35
Butyraldehyde-Propyl acetate	323-333	28	0.0030	0.02	0.0041	0.31

Group-interaction parameters:

$$a_{CCOO,CHO} = -49.60 \text{ K}$$
$$a_{CHO,CCOO} = 124.9 \text{ K}$$

TABLE VII

UNIQUAC parameters for the system 3Cℓ3F-ethane-n-Hexane

$r_1 = 4.0461$ $q_1 = 3.564$

$r_2 = 4.4998$ $q_2 = 3.856$

$a_{12} = 37.30$ K $a_{21} = -7.290$ K

Average deviations (both isotherms correlated simultaneously)

$|\Delta x| = 0.0033$ $|\Delta T| = 0.03$ K

$|\Delta y| = 0.0047$ $|\Delta P| = 0.05$ mmHg

VAPOR-LIQUID EQUILIBRIUM MEASUREMENTS ON FOUR BINARY SYSTEMS OF INDUSTRIAL INTEREST

W. Vincent Wilding, Loren C. Wilson and Grant M. Wilson ■ Wiltec Research Co., Inc., 488 S. 500 W., Provo, UT

Vapor-liquid equilibrium measurements have been made on four binary systems by the PTx method. Vapor and liquid phase compositions were derived from these data using a modified Redlich-Kwong equation of state to model the vapor phase and the Wilson equation to calculate liquid-phase activity coefficients.

The measurement of vapor-liquid equilibrium (VLE) data is one of the primary objectives of Project 805 of the Design Institute for Physical Property Data of the American Institute of Chemical Engineers. These data are of industrial importance and also contribute new information for correlative efforts in phase equilibria.

Project 805/84 consists of VLE measurements by the PTx method on four binary systems at two temperatures for each system. The following table lists the systems studied and the isotherms at which each system was studied.

Project 805/84

System	Isotherms,	°C
1) Ethyl Acetate/ Trimethylamine	0	90
2) tert-Butyl Acetate/ Isobutane	20	60
3) Dichloroethane/HCl	-30	0
4) DMF/1-Butene	20	90

APPARATUS AND PROCEDURE

The total pressure (PTx) method is a fast efficient method for the determination of binary vapor-liquid equilibria. The required measurements are total pressure versus charge composition at constant temperature and cell volume. The PTx method has the advantage that no vapor or liquid samples are taken, thus eliminating sampling and analyzing problems.

All four binary systems were studied using the agitated static cell shown in Figure 1. The apparatus consisted of a 300 cc stainless steel cell in an isothermal bath. The cell was equipped with charging, degassing, and pressure measurement lines. For pressures below approximately 100 kPa, a system of manometers was used to measure the pressure. The lines and first manometer were heated to a temperature higher than the bath to avoid condensation of the vapor outside of the cell. For pressures above 100 kPa, calibrated precision pressure gauges (3D Instruments, Inc.) were used. During a run, the cell was submerged in the constant temperature bath and manually agitated to assure equilibrium. The temperature was measured with a calibrated platinum resistance probe.

Each of the four binary systems was studied along two isotherms across the entire composition range. Each isotherm was traversed in two parts. First, the cell was charged with a known amount of one component. The cell was then degassed until a repeatable pure component vapor pressure was measured. Increments of the other component were then added to the cell and after the contents of the cell were degassed and allowed to come to equilibrium the pressure was measured. The second part of the procedure was similar to the first except that the second component was charged to the cell initially and increments of the first component were added. The range of mole fractions covered by the two halves were designed to overlap in the mid-composition region to check for consistency between the two parts.

The cell was degassed by weighing an evacuated cylinder or flask before and after a vapor sample was withdrawn. Additions were made using a weighed charging cylinder, syringe, or a positive displacement pump. The amount of material removed by degassing was accounted for in the data reduction procedure.

The PTx data reduction procedure is outlined in Gillespie et al. (1). Critical constants and other pure component data were obtained from Reid, Prausnitz, and Sherwood (2).

RESULTS AND DISCUSSION

The results of the experimental work are summarized in Tables 1 through 8. Each of the tables reporting PTx results list the run half, the concentration in the feed and in each phase, the measured and calculated pressure, the activity coefficient of each component, and the relative volatility. The parameters of the Wilson equation used to reduce the data are given at the bottom of each table.

Tables 1 and 2 present the results for the ethyl acetate/ trimethylamine system at 0°C and 90°C, respectively. Figure 2 shows the relative volatility of trimethylamine /ethyl acetate obtained from the reduced data.

The system of tert-butyl acetate and isobutane was studied at 30°C and 90°C. These results are summarized in Tables 3 and 4. Figure 3 presents the relative volatility of this system.

Tables 5 and 6 give the vapor-liquid equilibrium measurements for the dichloroethane/HCl system which were obtained at -30°C and 0°C. The relative volatility of HCl /dichloroethane is given in Figure 4.

Tables 7 and 8 show the results of PTx measurements on the DMF/1-butene system at 20°C and 90°C, respectively. Three Wilson equation parameters were required to obtain a good fit of the total pressure data because of the strong nonideality exhibited by this system. This modified Wilson equation is given in Wilding et al. (3). The relative volatility for this system is shown in Figure 5.

Table 9 is a comparison of measured and literature pure-component vapor pressures. The average error is 1.7% for the values shown. This good agreement with literature vapor pressures and the internal consistency of the measured data as evidenced by the overlap of run halves in the mid-composition region for each system indicates the reliability of the results of this research.

SUMMARY AND CONCLUSION

Vapor-liquid equilibrium measurements have been obtained by the PTx method on four binary systems. The results of this study will be useful for design purposes as well as for correlative endeavors.

We express gratitude to the Design Institute for Physical Property Data of the AIChE for sponsoring this research.

LITERATURE CITED

1. Gillespie, P.C., J.R. Cunningham, and G.M. Wilson, "Total Pressure and Infinite Dilution Vapor Liquid Equilibrium Measurements for the Ethylene Oxide/ Water System," AIChE Symposium Series. Experimental Results from the Design Institute for Physical Property Data I. Phase Equilibria, No. 244, Vol. 81, 26 (1985).

2. Reid, R.C., J.M. Prausnitz, and T.K. Sherwood, The Properties of Gases and Liquids, 3rd Ed. McGraw-Hill Book Co., New York, NY (1977).

3. Wilding, W.V., L.C. Wilson, and G.M. Wilson, "Vapor-Liquid Equilibrium Measurements on Ten Binary Systems of Industrial Interest", this volume.

4. Wichterle, I., J. LInek, E. Hála, Vapor-Liquid Equilibrium Data Bibliography, Elsevier Scientific Publishing Co. (1973).

Table 1. Ethyl Acetate/Trimethylamine Vapor-Liquid Equilibrium Measurements by PTx Method at 0.0°C.

Run	Mole Percent EAC			Pressure, kPa		Activity Coefficient		Relative Volatility
Half	Feed	Liquid	Vapor	Meas	Calc	EAC	TMA	TMA/EAC
2	0.00	0.00	0.00	89.30	89.30	1.616	1.000	15.58
2	3.35	3.41	0.22	86.52	86.41	1.545	1.001	16.32
2	6.37	6.54	0.41	84.20	83.87	1.488	1.003	17.00
2	9.66	9.97	0.62	81.73	81.16	1.431	1.006	17.74
2	19.70	20.36	1.26	74.48	73.34	1.296	1.024	19.98
2	29.76	30.74	1.96	67.34	65.76	1.200	1.051	22.20
2	44.94	46.27	3.27	55.46	54.22	1.104	1.107	25.49
1	44.96	45.56	3.20	54.99	54.76	1.107	1.104	25.34
1	54.81	55.55	4.36	47.01	46.91	1.066	1.148	27.42
2	55.34	56.94	4.55	46.04	45.78	1.061	1.155	27.71
1	69.38	70.07	7.15	34.84	34.48	1.027	1.222	30.40
1	78.87	79.35	10.64	26.84	25.72	1.012	1.276	32.27
1	89.40	89.70	20.24	14.80	15.04	1.003	1.341	34.33
1	94.76	94.85	34.28	9.32	9.34	1.001	1.376	35.34
1	97.67	97.71	54.30	5.90	6.06	1.000	1.395	35.90
1	100.00	100.00	100.00	3.36	3.36	1.000	1.412	-

WILSON PARAMETERS A = 0.560 B = 1.100 C = 1.000

Table 2. Ethyl Acetate/Trimethylamine Vapor-Liquid Equilibrium Measurements by PTx Method at 90.0°C.

Run	Mole Percent EAC			Pressure, kPa		Activity Coefficient		Relative Volatility
Half	Feed	Liquid	Vapor	Meas	Calc	EAC	TMA	TMA/EAC
2	0.00	0.00	0.00	1159.69	1159.69	1.392	1.000	3.87
2	3.38	3.62	0.93	1122.46	1119.31	1.353	1.001	4.01
2	6.48	6.90	1.76	1089.71	1083.86	1.321	1.002	4.14
2	9.89	10.47	2.66	1058.34	1046.24	1.289	1.004	4.28
2	20.23	21.06	5.38	948.37	939.92	1.208	1.016	4.69
2	30.61	31.44	8.23	844.60	840.47	1.146	1.035	5.11
2	45.52	45.60	12.83	714.98	707.67	1.085	1.072	5.70
1	46.16	46.71	13.24	700.50	697.35	1.081	1.075	5.74
1	55.47	55.80	17.07	610.53	611.53	1.053	1.105	6.13
2	56.87	57.12	17.71	599.84	598.92	1.050	1.109	6.19
1	69.95	70.52	26.09	468.15	468.37	1.022	1.163	6.78
1	79.24	79.83	35.48	373.07	373.74	1.010	1.205	7.20
1	89.62	90.05	54.11	263.17	264.95	1.002	1.258	7.67
1	94.81	95.05	70.84	201.12	209.49	1.001	1.286	7.91
1	97.69	97.81	84.71	174.44	178.36	1.000	1.301	8.04
1	100.00	100.00	100.00	153.20	153.20	1.000	1.314	-

WILSON PARAMETERS A = 0.697 B = 1.030 C = 1.000

Table 3. Tert-Butyl Acetate/Isobutane Vapor-Liquid Equilibrium Measurements by PTx Method at 20.0°C.

Run	Mole Percent TBA			Pressure, kPa		Activity Coefficient		Relative Volatility
Half	Feed	Liquid	Vapor	Meas	Calc	TBA	Isobutane	Isobutane/TBA
0	0.00	0.00	0.00	304.75	304.75	2.528	1.000	23.60
2	2.55	2.63	0.10	298.13	296.62	2.334	1.001	25.66
2	4.78	4.99	0.19	292.13	289.80	2.185	1.004	27.56
2	9.33	10.33	0.36	277.79	275.77	1.916	1.015	31.96
2	17.37	21.05	0.64	252.00	250.88	1.562	1.053	41.15
2	26.64	34.83	0.99	221.87	220.69	1.303	1.129	53.52
1	32.41	34.72	0.99	224.84	220.94	1.305	1.128	53.42
2	40.01	55.57	1.68	171.82	170.03	1.109	1.286	73.17
1	47.70	50.32	1.47	184.98	183.91	1.144	1.242	68.08
2	49.02	65.98	2.27	140.79	139.58	1.058	1.384	83.53
1	55.94	59.37	1.86	159.54	159.39	1.087	1.321	76.91
1	66.07	69.50	2.55	127.62	128.29	1.045	1.419	87.10
1	76.53	79.85	3.89	90.79	91.77	1.018	1.532	97.89
1	88.37	90.19	7.77	49.55	49.72	1.004	1.657	109.12
1	93.59	94.66	13.46	29.55	29.65	1.001	1.715	114.12
1	97.42	97.90	28.40	14.95	14.37	1.000	1.759	117.81
2	100.00	100.00	100.00	5.10	4.14	1.000	1.787	0.00
1	100.00	100.00	100.00	4.14	4.14	1.000	1.787	-

WILSON PARAMETERS A = 0.380 B = 1.040 C = 1.000

Table 4. Tert-Butyl Acetate/Isobutane Vapor-Liquid Equilibrium Measurements by PTx Method at 60.0°C.

Run	Mole Percent TBA			Pressure, kPa		Activity Coefficient		Relative Volatility
Half	Feed	Liquid	Vapor	Meas	Calc	TBA	Isobutane	Isobutane/TBA
2	0.00	0.00	0.00	873.90	873.84	2.394	1.000	8.67
2	1.98	2.17	0.24	854.25	853.21	2.247	1.001	9.31
2	5.78	6.50	0.65	818.75	815.55	2.007	1.006	10.61
2	10.07	11.36	1.05	782.55	777.40	1.800	1.017	12.11
2	20.58	22.90	1.84	695.68	696.36	1.474	1.059	15.83
2	30.21	33.28	2.52	623.63	627.25	1.298	1.112	19.33
1	44.11	46.22	3.47	535.72	537.58	1.162	1.196	23.89
2	44.53	48.06	3.63	514.69	524.18	1.148	1.209	24.56
2	54.64	58.76	4.75	437.12	441.65	1.083	1.292	28.55
1	58.86	62.26	5.23	414.03	412.75	1.067	1.322	29.89
1	68.54	72.57	7.22	329.98	321.32	1.032	1.415	33.99
1	79.16	83.21	11.42	229.39	216.27	1.011	1.521	38.44
1	89.44	90.63	18.83	135.00	136.20	1.003	1.600	41.69
1	94.64	95.28	31.56	83.01	83.00	1.001	1.653	43.80
1	97.55	97.81	49.82	53.78	53.14	1.000	1.682	44.97
1	100.00	100.00	100.00	26.68	26.68	1.000	1.708	-

WILSON PARAMETERS A = 0.384 B = 1.084 C = 1.000

Table 5. Dichloroethane/HCl Vapor-Liquid Equilibrium by PTx Method at -30.0°C.

Run Half	Mole Percent DCE Feed	Mole Percent DCE Liquid	Mole Percent DCE Vapor	Pressure, kPa Meas	Pressure, kPa Calc	Activity Coefficient DCE	Activity Coefficient HCl	Relative Volatility HCl/DCE
2	0.00	0.00	0.00	1097.64	1097.64	0.610	1.000	2880.54
2	3.06	3.29	0.00	1067.99	1055.33	0.636	0.999	2808.48
2	4.55	4.88	0.00	1047.30	1034.48	0.649	0.998	2775.69
2	9.17	9.73	0.00	979.74	969.78	0.686	0.994	2684.81
2	19.65	20.48	0.01	831.50	824.21	0.761	0.976	2517.05
2	30.71	31.50	0.02	682.44	677.62	0.828	0.948	2379.71
1	45.61	45.73	0.04	500.69	500.26	0.897	0.902	2234.06
2	46.39	46.84	0.04	491.73	487.25	0.901	0.898	2223.84
1	54.05	54.26	0.05	396.86	402.54	0.929	0.871	2158.10
2	56.48	56.66	0.06	384.72	376.32	0.937	0.861	2137.88
1	70.29	70.59	0.12	230.01	235.59	0.972	0.807	2028.57
1	79.09	79.36	0.20	156.10	156.89	0.987	0.772	1965.63
1	89.69	89.87	0.47	72.39	72.36	0.997	0.730	1895.00
1	93.58	93.70	0.79	44.47	44.08	0.999	0.715	1870.35
1	95.79	95.87	1.23	28.82	28.63	1.000	0.707	1856.66
1	100.00	100.00	100.00	0.36	0.36	1.000	0.691	-

WILSON PARAMETERS A = 0.900 B = 1.600 C = 1.000

Table 6. Dichloroethane/HCl Vapor-Liquid Equilibrium by PTx Method at 0.0°C.

Run Half	Mole Percent DCE Feed	Mole Percent DCE Liquid	Mole Percent DCE Vapor	Pressure, kPa Meas	Pressure, kPa Calc	Activity Coefficient DCE	Activity Coefficient HCl	Relative Volatility HCl/DCE
2	0.00	0.00	0.00	2616.88	2616.71	0.717	1.000	478.87
2	3.03	3.65	0.01	2500.71	2488.01	0.735	1.000	490.60
2	4.48	5.39	0.01	2439.14	2427.31	0.744	0.999	496.08
2	8.97	10.54	0.02	2271.25	2249.41	0.768	0.996	511.00
2	18.82	21.51	0.05	1902.39	1883.08	0.818	0.984	537.74
2	29.13	32.39	0.09	1553.72	1540.69	0.863	0.965	558.02
1	43.45	45.79	0.15	1168.65	1152.33	0.910	0.933	574.33
2	44.39	47.26	0.16	1114.87	1112.09	0.915	0.928	575.56
1	51.22	54.42	0.21	925.96	923.15	0.936	0.907	579.98
2	54.25	56.79	0.23	871.28	863.23	0.942	0.899	580.90
1	67.43	70.85	0.42	533.30	535.10	0.974	0.849	580.99
1	77.12	79.61	0.67	354.18	354.34	0.987	0.814	576.77
1	88.32	90.03	1.56	162.51	162.71	0.997	0.770	567.91
1	92.70	93.80	2.62	97.63	99.42	0.999	0.754	563.76
1	95.45	95.94	4.04	68.88	65.08	0.999	0.744	561.21
1	100.00	100.00	100.00	2.68	2.68	1.000	0.726	-

WILSON PARAMETERS A = 1.125 B = 1.215 C = 1.000

Table 7. DMF/1-Butene Vapor-Liquid Equilibrium Measurements by PTx Method at 20.0°C.

Run Half	Mole Percent DMF Feed	Liquid	Vapor	Pressure, kPa Meas	Calc	Activity Coefficient DMF	Butene	Relative Volatility Butene/DMF
2	0.00	0.00	0.00	255.72	255.72	18.422	1.000	30.93
2	2.73	2.81	0.06	250.69	249.48	12.848	1.005	44.68
2	5.06	5.26	0.09	247.24	245.65	9.878	1.016	58.85
2	10.16	10.65	0.12	243.24	240.59	6.250	1.056	96.92
2	19.76	20.80	0.13	238.63	237.37	3.462	1.177	195.19
2	29.32	30.90	0.14	237.04	237.05	2.340	1.347	330.55
1	43.18	44.73	0.14	234.63	235.31	1.633	1.673	588.68
2	43.79	45.92	0.14	232.83	234.90	1.594	1.707	615.46
2	53.76	56.07	0.14	228.08	228.22	1.339	2.045	880.53
1	53.88	55.95	0.14	224.56	228.34	1.341	2.041	876.86
1	67.38	70.16	0.17	204.22	203.52	1.136	2.704	1385.32
1	76.74	79.17	0.21	172.44	171.40	1.063	3.292	1826.81
1	88.65	90.46	0.37	100.80	100.77	1.013	4.307	2578.84
1	93.28	94.54	0.59	63.37	63.42	1.004	4.779	2926.33
1	96.92	97.28	1.11	31.00	33.82	1.001	5.135	3189.17
1	100.00	100.00	100.00	0.38	0.38	1.000	5.523	-

WILSON PARAMETERS A = 0.171 B = 0.670 C = 1.390

Table 8. DMF/1-Butene Vapor-Liquid Equilibrium Measurements by PTx Method at 90.0°C.

Run Half	Mole Percent DMF Feed	Liquid	Vapor	Pressure, kPa Meas	Calc	Activity Coefficient DMF	Butene	Relative Volatility Butene/DMF
2	0.00	0.00	0.00	1475.81	1475.81	10.235	1.000	5.10
2	2.77	2.95	0.45	1437.54	1431.20	7.844	1.004	6.78
2	5.19	5.54	0.69	1407.55	1400.25	6.410	1.013	8.46
2	10.52	11.16	0.98	1358.95	1351.18	4.466	1.046	12.75
2	20.58	21.57	1.18	1296.89	1297.21	2.760	1.148	23.04
2	30.64	31.74	1.27	1258.63	1260.41	1.998	1.289	36.20
1	44.63	44.85	1.37	1201.68	1201.57	1.501	1.537	58.58
2	45.70	46.50	1.38	1185.20	1191.65	1.459	1.574	61.91
2	55.66	56.70	1.50	1102.81	1109.72	1.261	1.838	85.81
1	55.93	56.28	1.50	1116.25	1113.91	1.268	1.826	84.70
1	69.60	71.58	1.84	919.75	899.73	1.097	2.354	134.44
1	78.51	80.67	2.32	713.95	697.66	1.042	2.772	175.94
1	89.78	91.31	4.17	357.01	369.08	1.008	3.396	241.63
1	94.07	95.08	6.66	226.01	225.78	1.003	3.660	270.58
1	97.01	97.56	12.07	122.31	122.56	1.001	3.849	291.66
1	100.00	100.00	100.00	14.52	14.52	1.000	4.047	-

WILSON PARAMETERS A = 0.228 B = 0.780 C = 1.370

Table 9. Comparison of Pure Component Vapor Pressures to Literature Values.

System	Compound	Temp (°C)	Vapor Pressure (kPa) Measured	Literature	Source
1	Ethyl Acetate	0	3.36	3.37	a
		90	153.20	151.01	a
	Trimethylamine	0	89.30	90.27	a
		90	1159.69	1274.90	a
2	Tert-butyl Acetate	20	4.14	-	
		60	26.68	-	
	Isobutane	20	304.75	300.34	a
		60	873.90	851.80	a
3	Dichlorethane	-30	0.36	0.364	a
		0	2.68	2.80	a
	HCl	-30	1097.64	1089.41	a
		0	2616.88	2598.30	a
4	DMF	20	0.38	0.377	b
		90	14.52	14.52	b
	1-Butene	20	255.72	254.65	a
		90	1475.81	1450.45	a

(a) Antoine equation from Reference 2.
(b) Antoine equation from Reference 4.

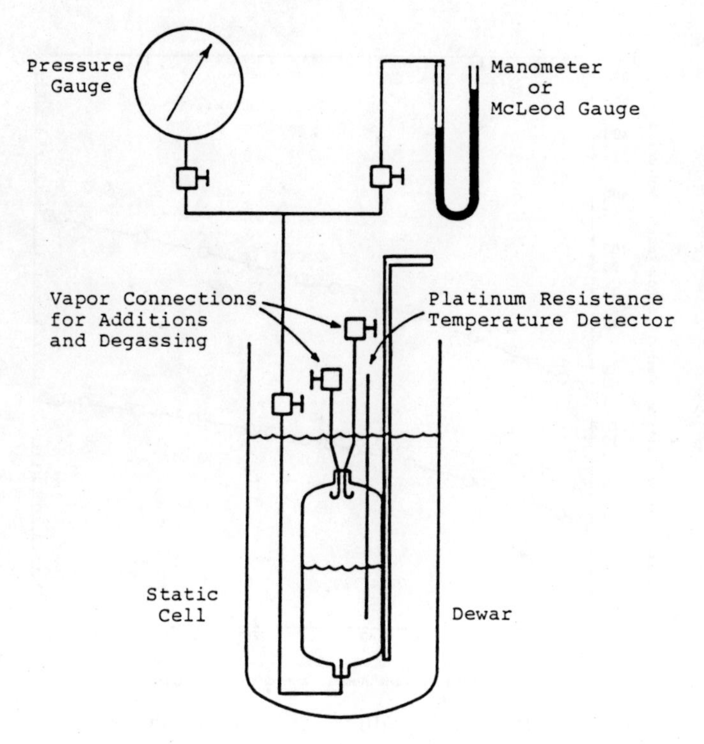

Figure 1. Schematic of agitated static cell used for PTx measurements.

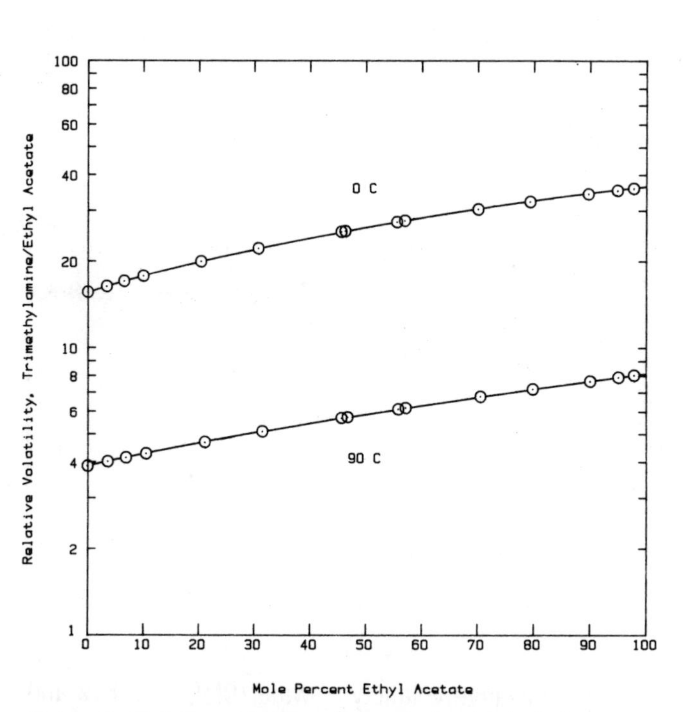

Figure 2. Relative volatility of trimethylamine/ethyl acetate from PTx data.

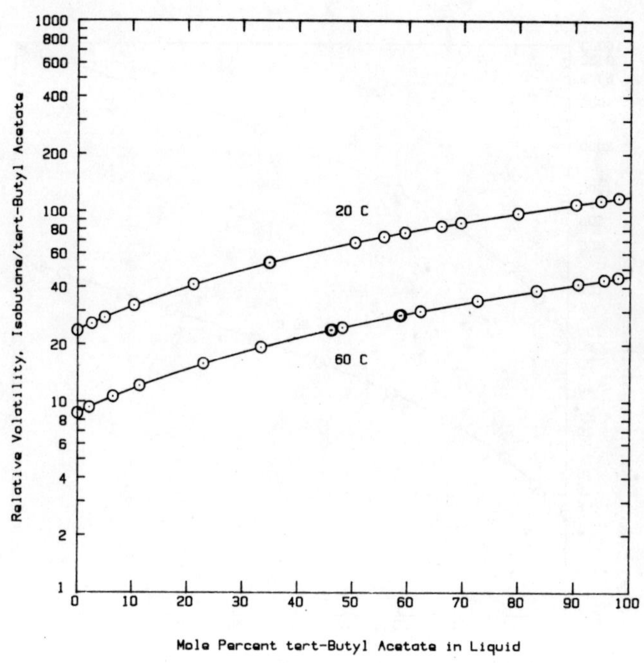

Figure 3. Relative volatility of isobutane/tert-butyl acetate from PTx data.

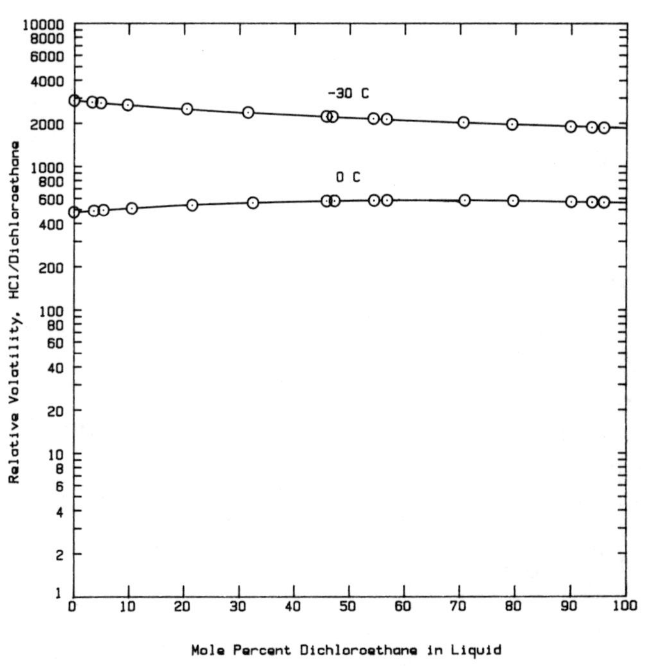

Figure 4. Relative volatility of HC1/dichloroethane from PTx data.

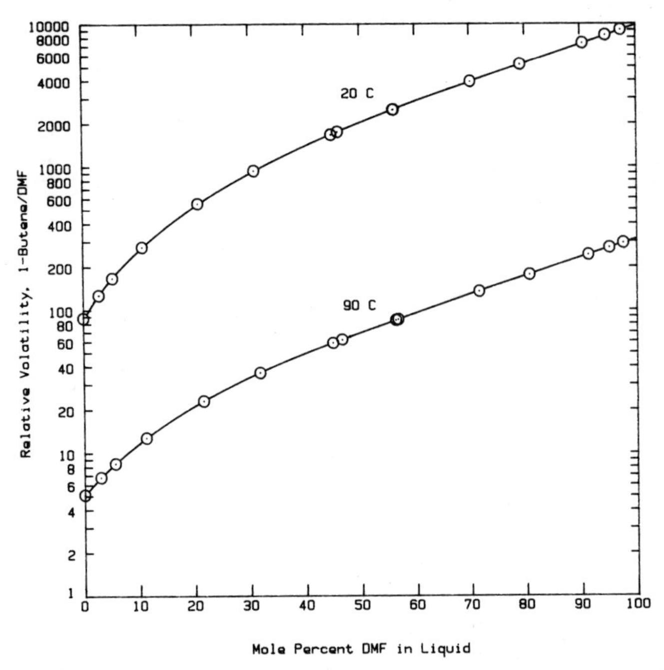

Figure 5. Relative volatility of 1-butene/DMF from PTx data.

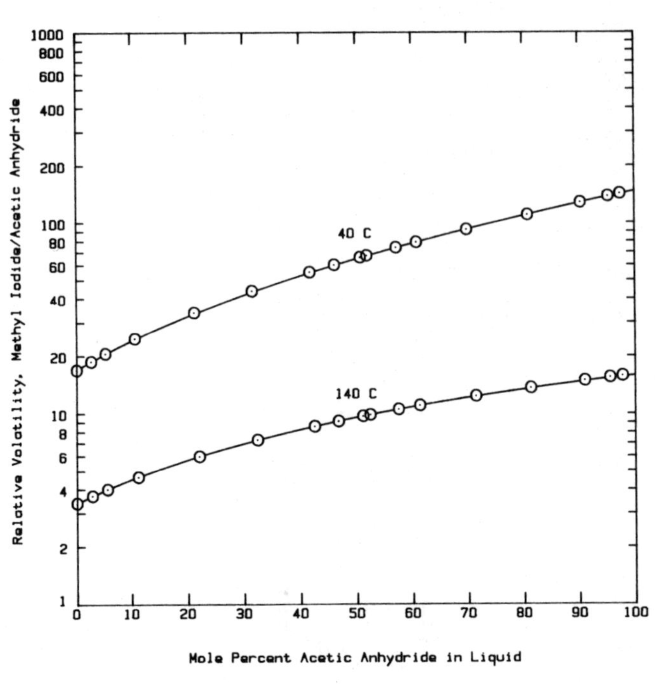

Figure 7. Relative volatility of methyl iodide/acetic anhydride from PTx data.

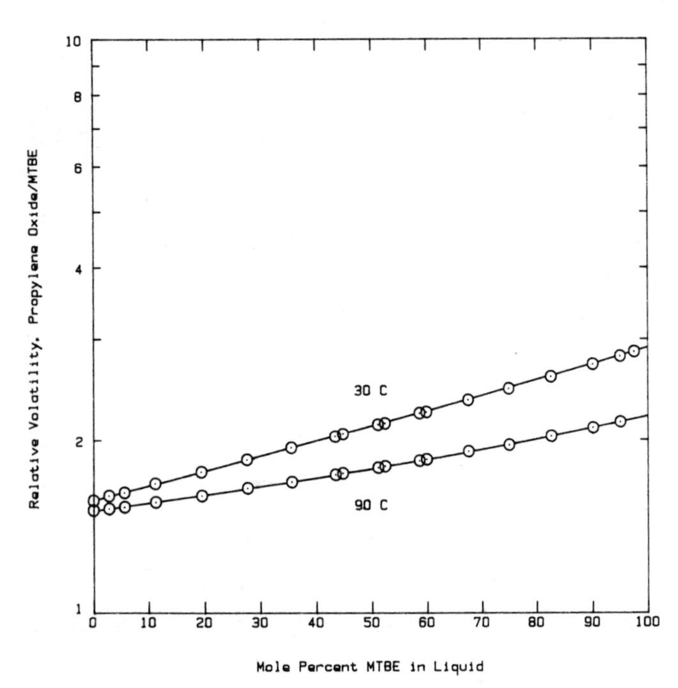

Figure 6. Relative volatility of propylene oxide/methyl tert-butyl ether from PTx data.

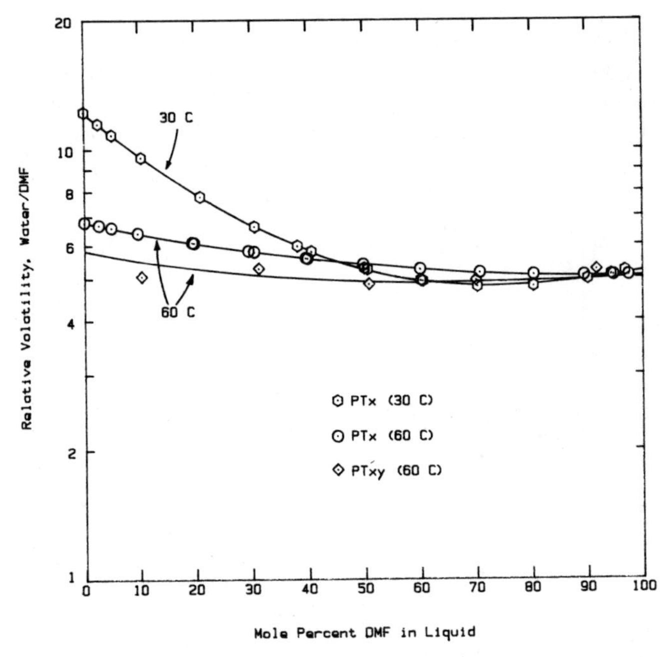

Figure 8. Relative volatility of water/DMF from PTx and PTxy data.

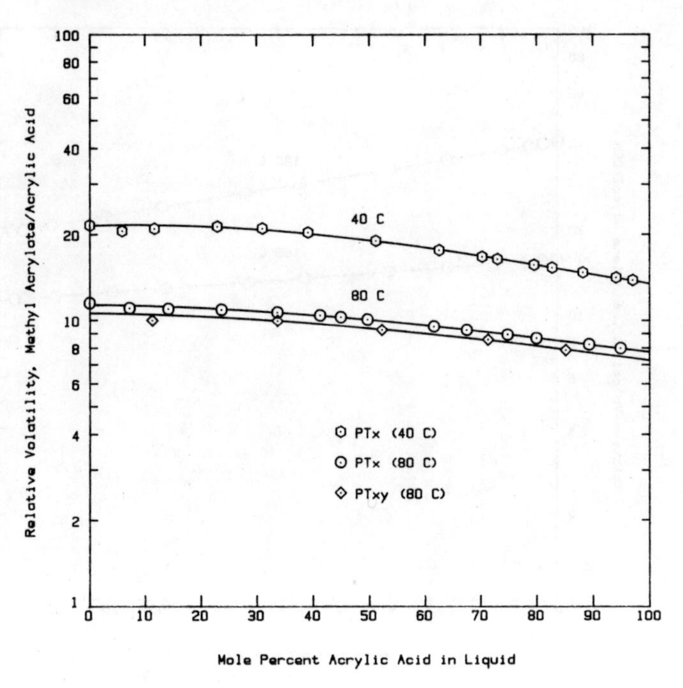

Figure 9. Relative volatility of methyl acrylate/acrylic acid from PTx and PTxy data.

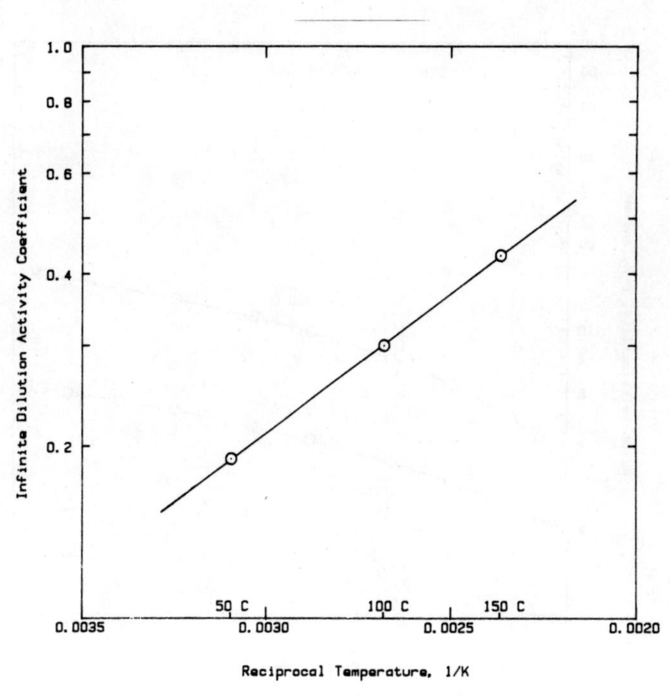

Figure 11. Infinite dilution activity coefficient of chloroform in NMP by gas chromatography.

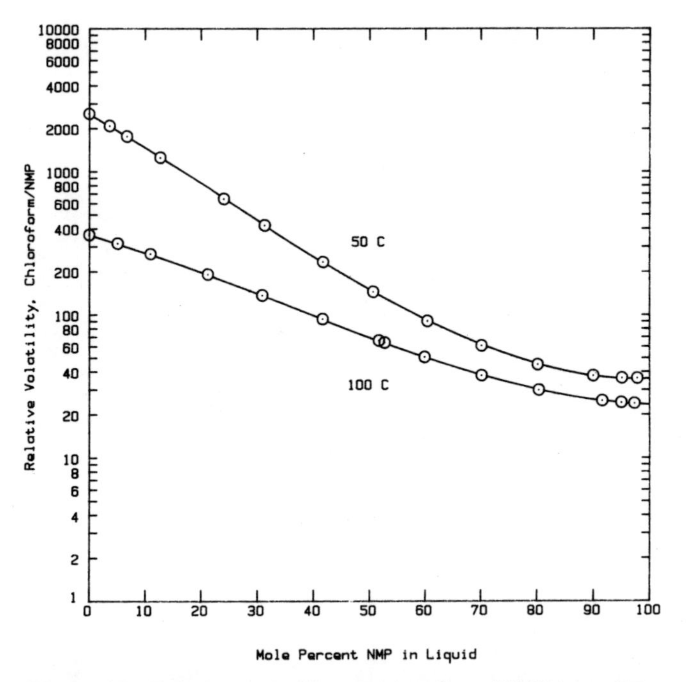

Figure 10. Relative volatility of chloroform/NMP from PTx data.

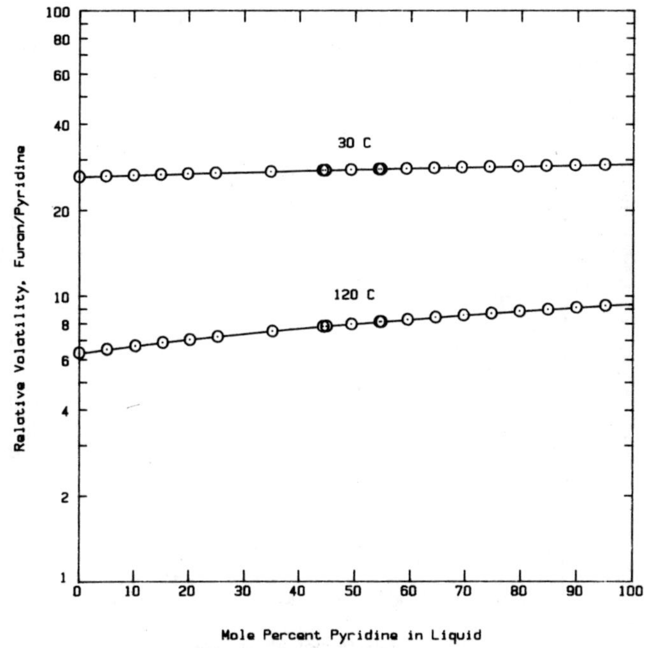

Figure 12. Relative volatility of furan/pyridine from PTx data.

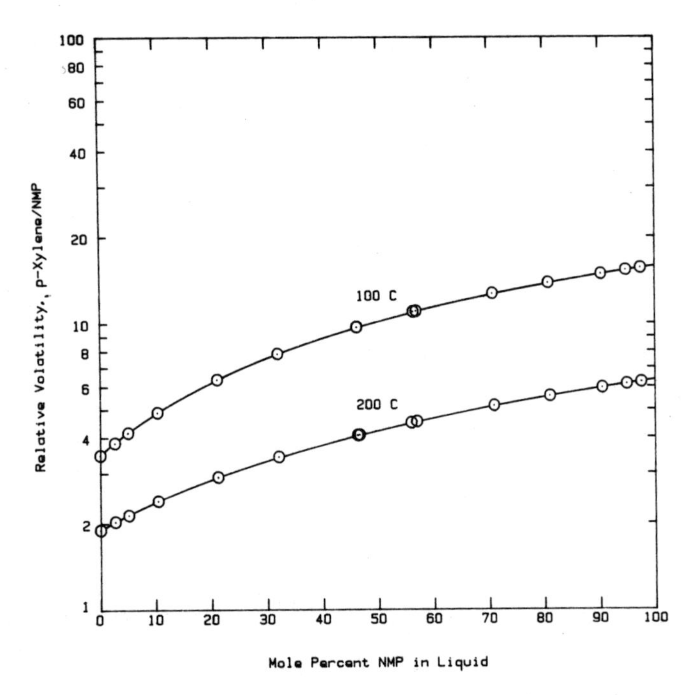

Figure 13. Relative volatility of p-xylene/NMP from PTx data.

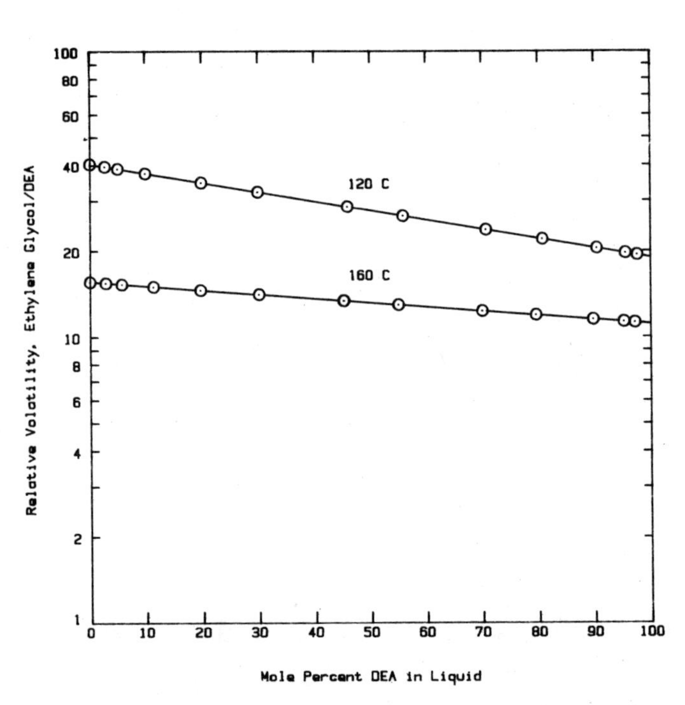

Figure 15. Relative volatility of ethylene glycol/diethanolamine from PTx data.

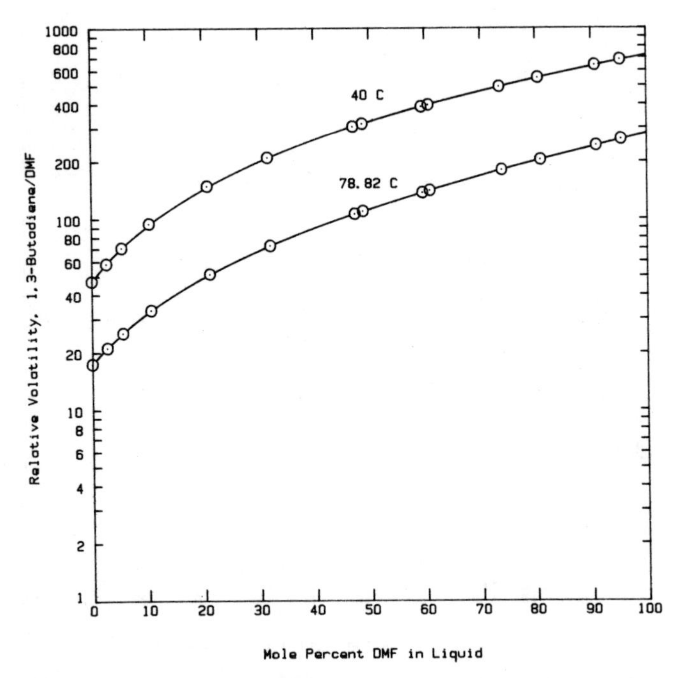

Figure 14. Relative volatility of 1,3-butadiene/DMF from PTx data.

EXPERIMENTAL VAPOR LIQUID EQUILIBRIA FOR THE ACETONITRILE/DIETHYLAMINE SYSTEM

J.W. Blasdel, B.E. Poling and D.B. Manley ■ The University of Missouri-Rolla, Rolla, MO

Isothermal pressure, liquid composition, vapor composition (TPXY) measurements were made for the acetonitrile / diethylamine system at two temperatures. The results are given in Table 1 and shown in Figures 2 and 3.

The pure chemicals were obtained from Aldrich Chemical Company. The diethylamine was distilled at atmospheric pressure but this did not improve the purity. Table 2 shows the gas chromatographic analyses before and after distillation.

Figure 1 shows the recirculating glass still designed by David Zudkevitch and purchased from Ace Glass Company which was used for the experiments. A discussion of these types of stills is given on pages 283-299 of Reference 1. The system pressure was measured with a U-tube mercury manometer, and the vapor and liquid temperatures were measured with calibrated mercury thermometers.

A mixture of the two chemicals was circulated in the still at the specified liquid temperature until a steady state was well established. The vapor temperature and circulation rate were then carefully adjusted to assure near equilibrium conditions. Vapor and liquid samples were withdrawn by syringe.

The samples were injected into a Beckman gas chromatograph equipped with a flame ionization detector and a Supelco 1/8" by 6' stainless steel column packed with 5% SP-2100 on 100/200 supelcoport. The gas chromatograph was calibrated with seven mixtures made up by weight to cover the entire composition range. The ratio of peak area fraction to mole fraction was judged to be independent of composition within experimental error. From three to five injections were made for each calibration sample and each still sample.

The experimental data in Table 1 were correlated with the Antoine vapor pressure equation and the UNIQUAC activity coefficient correlation with temperature dependent parameters. The regression was global, and each experimental data value was weighted by its estimated experimental uncertainty given in Table 3. Values of the experimental variables calculated with the correlation are plotted in Figures 2 and 3. Pure component vapor pressures from the DIPPR data bank shown in the figures, match closely the experimental values of this study. Sources of other experimental values are given in References 2 and 3.

Reference 4 reports an azeotrope for the acetonitrile/diethylamine system at 1 atm, 54.5^{o}C, and 87 mole % diethylamine. This is consistent with the values in Table 1, and the azeotrope indicated in Figures 2 and 3.

Nothing unusual was observed during the course of the experiments.

REFERENCES

1. Hala, E., J. Pick, V. Fried, and O. Vilim, Vapour-Liquid Equilibrium, 2nd ed., Pergamon, 1967.

2. Boublik, T., V. Fried, and E. Hala, The Vapour Pressures of Pure Substances, 2nd ed., Elsevier, 1984.

3. T. E. Jordan, Vapor Pressure of Organic Compounds, Interscience Publishers, London, 1954.

4. CRC Handbook of Chemistry and Physics, 61st ed., page D-4, CRC Press, Inc. 1980-81.

TABLE 2
ANALYSIS OF PURE CHEMICALS
(mole percent)

Chemical	Packaged purity	Before purification	After purification
Diethylamine	98.0	99.87	99.87
Acetonitrile	99.0	99.57	

TABLE 3
EXPERIMENTAL UNCERTAINTIES

Measured quantity	Estimated uncertainty
Temperature	0.2 deg. C.
Pressure	1.0 mmHg
Liquid composition	1.0 mole percent
Vapor composition	2.0 mole percent

TABLE 1
ACETONITRILE (1) / DIETHYLAMINE (2)
EXPERIMENTAL RESULTS

P mmHg	X(2) mole frac.	Y(2) mole frac.
35.0 deg C		
135.5	0.0000	0.0000
155.1	0.0216	0.1482
159.5	0.0233	0.1550
199.9	0.0724	0.3664
205.5	0.0810	0.3919
276.4	0.2143	0.5765
279.6	0.2167	0.5802
331.9	0.4610	0.7288
332.0	0.4704	0.7197
354.4	0.7421	0.8264
353.4	0.7436	0.8301
360.6	0.8900	0.8879
359.5	0.8921	0.8891
360.4	0.9489	0.9304
361.6	0.9499	0.9288
352.2	1.0000	1.0000
50.0 deg C		
250.8	0.0000	0.0000
273.0	0.0137	0.0926
280.6	0.0163	0.1265
345.2	0.0388	0.3231
349.9	0.0639	0.3781
483.7	0.1605	0.5911
479.3	0.1623	0.5737
591.1	0.4346	0.7304
591.4	0.4509	0.7328
638.9	0.7106	0.8196
638.7	0.7185	0.8246
647.9	0.8909	0.8830
646.6	0.8912	0.8869
638.7	0.9501	0.9340
636.6	0.9507	0.9285
625.5	1.0000	1.0000

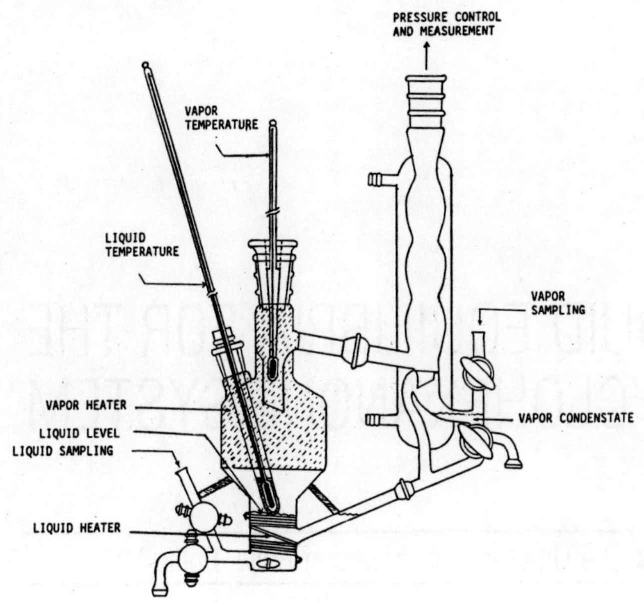

FIGURE 1.
VAPOR/LIQUID RECIRCULATING STILL

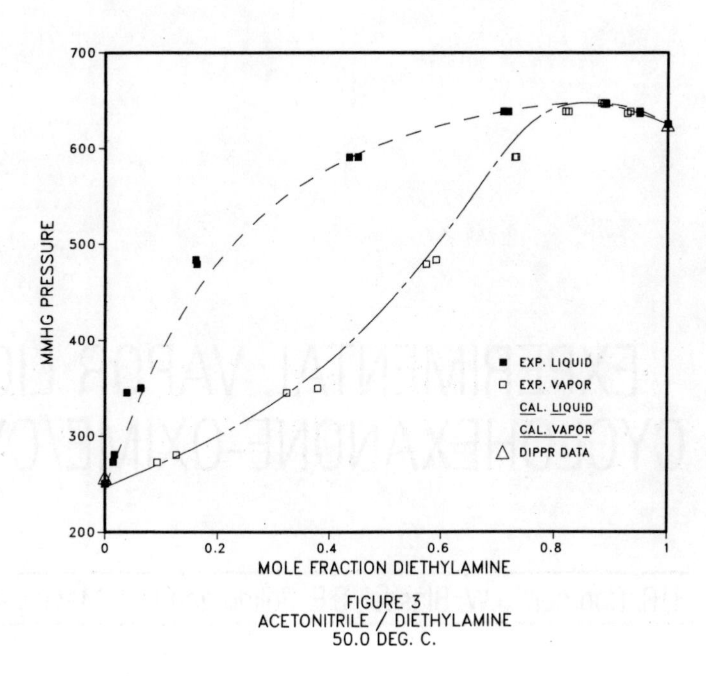

FIGURE 3
ACETONITRILE / DIETHYLAMINE
50.0 DEG. C.

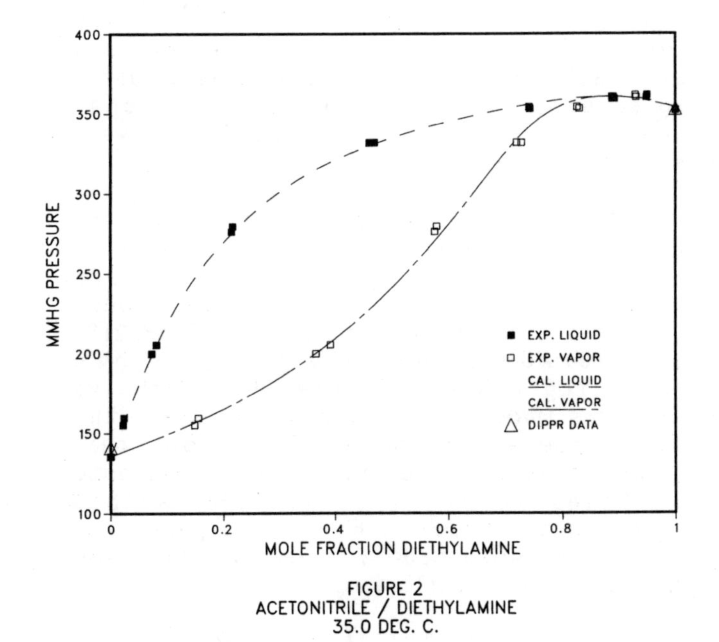

FIGURE 2
ACETONITRILE / DIETHYLAMINE
35.0 DEG. C.

EXPERIMENTAL VAPOR LIQUID EQUILIBRIA FOR THE CYCLOHEXANONE-OXIME/CYCLOHEXANONE SYSTEM

L.R. Dohmen, J.W. Blasdel, B.E. Poling and D.B. Manley ■ The University of Missouri-Rolla, Rolla, MO

Isothermal pressure, liquid composition, vapor composition (TPXY) measurements were made for the cyclohexanone-oxime / cyclohexanone system at 136.2°C and 151.6°C. The results are given in Table 1 and shown in Figures 2 and 3.

The pure chemicals were obtained from Aldrich Chemical Company and were further purified before use. Table 2 shows the gas chromatograph analyses before and after purification. The cyclohexanone-oxime was purified by sublimation under vacuum and collection on an ice-cold surface. The cyclohexanone was purified by pouring over molecular sieves and distillation in a nitrogen atmosphere at atmospheric pressure. Pure ethanol was used as a solvent for the gas chromatographic analysis.

Figure 1 shows the recirculating glass still designed by David Zudkevitch and purchased from Ace Glass Company which was used for the experiments. A discussion of these types of stills is given on pages 283-299 of Reference 1. A key feature of the still shown in Figure 1 is the heating jacket to prevent condensation of the vapor phase. This was particularly important for the compounds in this study due to large separation in the pure component boiling points, and due to the high temperatures involved. The system pressure was measured with a U-tube mercury manometer, and the vapor and liquid temperatures were measured with calibrated mercury thermometers.

A mixture of the two chemicals was circulated in the still at the specified liquid temperature until steady state was well established. The vapor temperature and circulation rate were then carefully adjusted to assure near equilibrium conditions. Samples of liquid and condensed vapor were withdrawn by syringe.

The samples were dissolved in ethanol and then injected into a Beckman gas chromatograph equipped with a flame ionization detector and a Supleco 1/8" by 6' stainless steel column packed with 5% SP-2100 on 100/200 supelcoport. The gas chromatograph was calibrated with twelve mixtures made up by weight to cover the entire composition range. The ratio of peak area fraction to mole fraction was judged to be independent of composition within experimental error. From three to five injections were made for each calibration sample and each still sample.

The experimental data in Table 1 were correlated with the Antoine vapor pressure equation and the UNIQUAC activity coefficient correlation with temperature dependent parameters. The regression was global, and each experimental data value was weighted by its estimated experimental uncertainty. Values of the experimental variables calculated with the correlation are plotted in Figures 2 and 3. Pure component vapor pressures from the DIPPR data bank also shown in the figures, are about 18mm lower than the values in this study. The reason for this difference is not known. The DIPPR values are consistent with the experimental values given in References 2 and 3.

The estimated experimental uncertainties in the measured quantities, which are listed in Table 3 are higher than normal, particularly for the vapor composition. There were several reasons for this. The composition measurements for high concentrations of cyclohexanone were relatively inaccurate due to overlapping of the peaks in the gas chromatograph analysis. However, the greatest source of error was due to the reactive nature of the chemicals. Some discoloration of the chemicals was observed in the still during the experiments, indicating that some reaction may have been occurring. Also, additional peaks appeared in the gas chromatograms when samples were rerun after being stored for several days at room temperature. Finally, it was extremely difficult to get reproduceable results from the gas chromatograph. Numerous columns, solvents, and operating conditions were tried before settling on the procedure used. Even then it was necessary to discard some chromatograms when it was obvious that anomalies were taking place in the gas chromatograph.

REFERENCES

1. Hala, E., J. Pick, V. Fried, and O. Vilim, Vapour-Liquid Equilibrium, 2nd ed., Pergamon, 1967.

2. Boublik, T., V. Fried, and E. Hala, The Vapour Pressures of Pure Substances, 2nd ed., page 464, Elsevier, 1984.

3. Stull, Ind. Eng. Chem. 39, 517 (1947).

TABLE 1
CYCLOHEXANONE OXIME (1) / CYCLOHEXANONE (2)
EXPERIMENTAL RESULTS

P mmHg	X(2) mole frac.	Y(2) mole frac.
	136.2 deg C	
57.9	0.0000	0.0000
71.5	0.0095	0.4167
91.3	0.0653	0.4399
150.7	0.2245	0.7174
270.6	0.4804	0.9606
366.7	0.7288	0.9779
402.6	0.8897	0.9929
440.2	0.9401	0.9890
468.7	1.0000	1.0000
	151.6 deg C	
105.9	0.0000	0.0000
132.5	0.0295	0.3876
158.3	0.0693	0.3266
261.7	0.2117	0.7042
413.2	0.4142	0.9554
413.4	0.4380	0.9549
412.9	0.4478	0.9561
439.2	0.4531	0.9482
427.1	0.4708	0.9580
560.2	0.7305	0.9876
645.0	0.8907	0.9904
666.8	0.9414	0.9951
706.8	1.0000	1.0000

TABLE 2
ANALYSIS OF PURE CHEMICALS
(mole percent)

Chemical	Pack-aged purity	Before purifica-tion	After purifica-tion
Cyclohexanone-oxime	97.0	98.30	99.87
Cyclohexanone	99.8	99.89	99.95
Ethanol		99.96	

TABLE 3
EXPERIMENTAL UNCERTAINTIES

Quantity	Uncertainty
Temperature	0.2 deg. C.
Pressure	1.0 mmHg
Liquid comp.	3.0 mole %
Vapor comp.	3.0 mole % (> 10% oxime)
Vapor comp.	10.0 mole % (< 10% oxime)

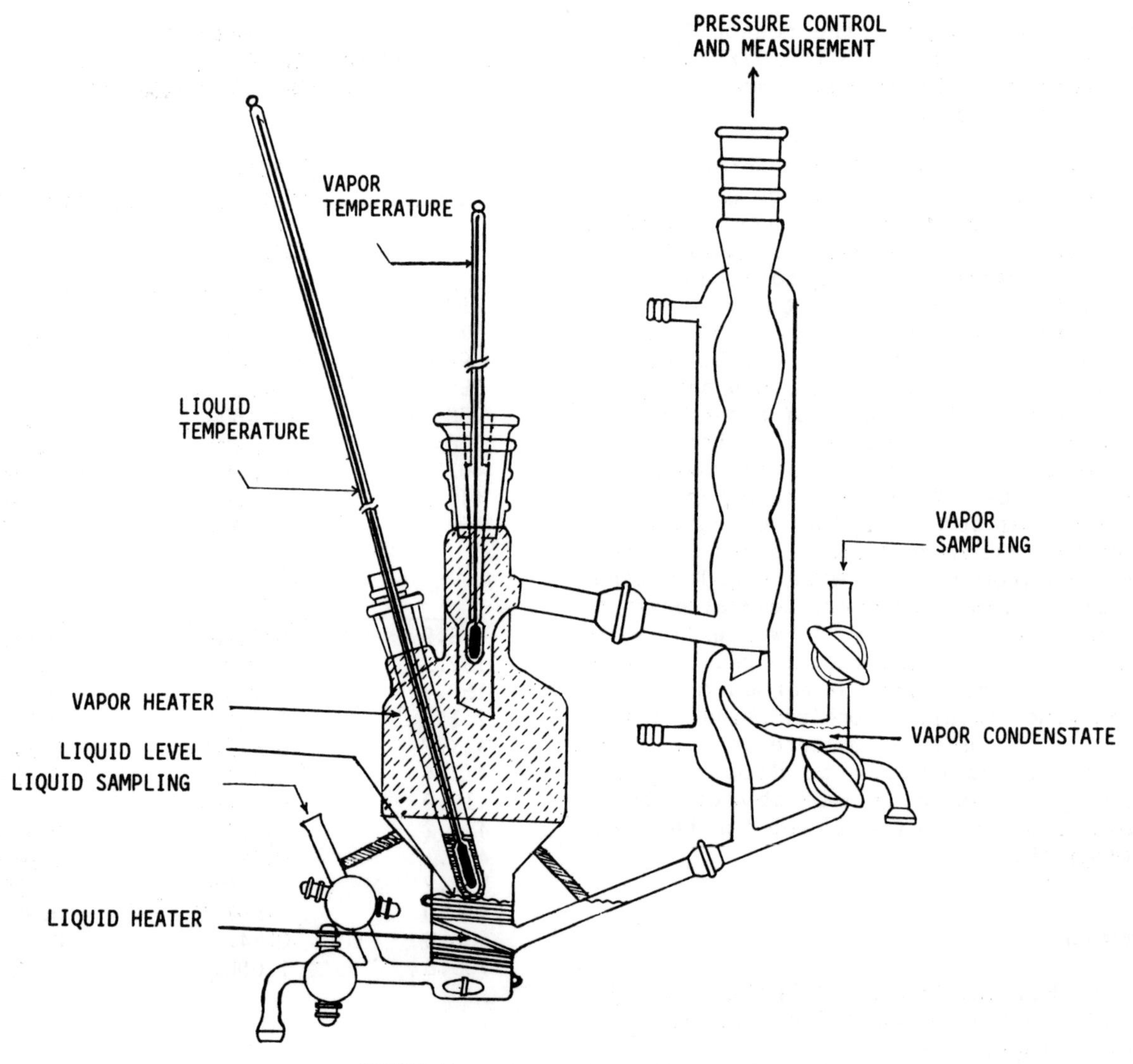

FIGURE 1. VAPOR/LIQUID RECIRCULATING STILL

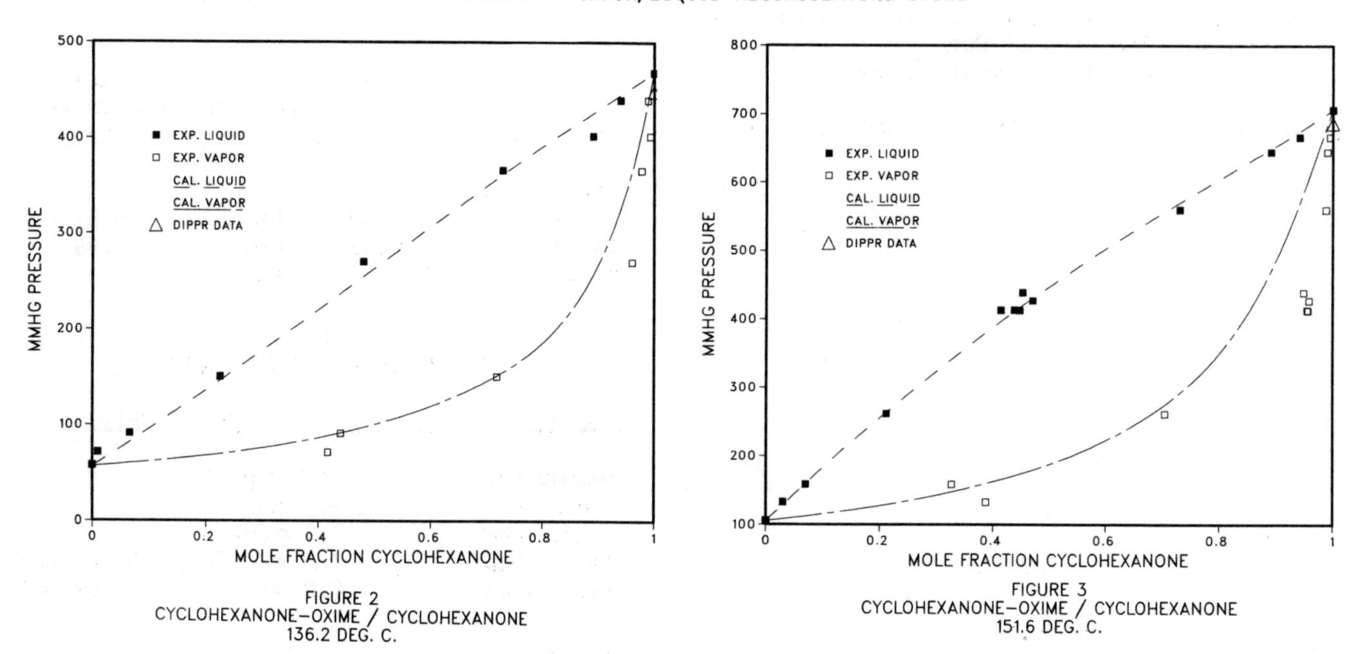

FIGURE 2
CYCLOHEXANONE–OXIME / CYCLOHEXANONE
136.2 DEG. C.

FIGURE 3
CYCLOHEXANONE–OXIME / CYCLOHEXANONE
151.6 DEG. C.

VAPOR-LIQUID EQUILIBRIUM MEASUREMENTS ON THE AMMONIA-WATER SYSTEM FROM 313 K TO 589 K

Paul C. Gillespie, W. Vincent Wilding and Grant M. Wilson ■ Wiltec Research Co., Inc., 488 S. 500 W., Provo, UT

Vapor-liquid equilibrium data for the ammonia-water system over the complete composition range have been obtained at eight temperatures from 313.15 K to 588.7 K. The total pressure method was used to obtain PTx data, and in a separate procedure equilibrium vapor and liquid phase compositions (PTxy data) were analyzed.

To evaluate the consistency of the two data sets, the PTx data were reduced to PTxy data using the Redlich-Kister activity coefficient equation and an equation of state containing a hard-sphere expansion term and the second and third virial coefficients. The parameters of the Redlich-Kister expansion were obtained by fitting the total pressure data using a least-squares procedure. One PTxy datum was used to adjust the value of the second cross virial coefficient. Following this procedure, relative volatilities calculated from the total pressure data were in good agreement with the values obtained directly from the equilibrium phase measurements. The result is a thermodynamically consistent set of vapor-liquid equilibrium data.

The ammonia-water binary system is important in many industrial processes. Optimum design and operation of these processes require accurate vapor-liquid equilibrium (VLE) data. Because of the nonideality of the ammonia-water system accurate data have not been plentiful.

The first attempt to obtain data on the ammonia-water binary over the full range of composition was made by T. A. Wilson (1) who measured total pressure and VLE data from 273.15 K to 364.15 K and to 1.170 MPa. His data were further extrapolated to 3.850 MPa. The extrapolated data are published in the Chemical Engineers' Handbook (2).

Wucherer (3) measured data from 273.15 K to 466.4 K and to pressures of 2.070 MPa. These data cover the full composition range along intermediate isotherms from 294.3 K to 322.0 K. Vapor phase compositions were presented to an accuracy of one part per thousand, which for low water concentrations resulted in only one significant figure; e.g., 0.2 wt percent water in the vapor phase.

Because of this low level of accuracy, Scatchard et al. (4) used the total pressure data of Wucherer to obtain the vapor composition. They calculated the chemical potential of water by performing a Gibbs-Duhem integration and assuming that water was essentially nonvolatile. The vapor concentrations of water predicted by Scatchard were much lower than the data of Wucherer in the low-water region. At low temperatures there is as much as an order of magnitude difference.

Macriss et al. (5) measured dew point pressures at low concentrations of water and at temperatures from 333.9 K to 389.6 K. Their data fall between the data of Wucherer and the calculated values of Scatchard et al. but are within the expressed experimental accuracy of Wucherer's results. Macriss et al. combined their data with the data of Wucherer giving a data set from 0.007 MPa to 3.450 MPa.

Edwards et al. (6) and Won et al. (7) calculated vapor compositions from total pressure data using a new activity coefficient equation and an equation of state for the vapor phase that attempts to account for the interaction of polar compounds. Their results are essentially the same as the results of Scatchard et al. These authors also concluded that the data of T. A. Wilson and the compilation of Macriss et al. are thermodynamically inconsistent stating that the reported vapor concentrations of water are too high.

More recently, Guillevic et al. (8) have published PTxy data at temperatures from 403.1 K to 503.1 K and to pressures of 7.0 MPa. The vapor and liquid compositions were determined by chromatographic analysis of micro-samples

Research sponsored jointly by the Design Institute for Physical Property Data of the American Institute of Chemical Engineers and the Gas Processors Association.

of the two equilibrium phases. Also, Rizvi (9) presents PTxy data in his Ph.D. thesis obtained from 306 K to 618 K and to pressures of 22 MPa. The phase compositions were obtained by analyzing small samples by gas chromatography as was done by Guillevic et al.

The discrepancies and gaps in existing data caused Bogart (10) to conclude, "Anomalies in the published data on the binary system, ammonia-water, have necessitated the use of mathematical correlation and extrapolation techniques to provide suitable coverage..."

In order to help resolve these discrepancies the present study was performed at Wiltec Research Company, Inc. under the joint sponsorship of The Gas Processors Association and the Design Institute for Physical Property Data of the American Institute of Chemical Engineers.

APPARATUS AND PROCEDURE

PTx Measurements

Total pressure (PTx) measurements were made in the rocked cell shown in Figure 1. The cell was made of Type 316 stainless steel with an inside volume of 312.4 cc. The cell was insulated and heated, and the temperature was measured to an accuracy of ±0.01 K with a platinum resistance temperature probe inserted in a well in the center of the cell. The probe was calibrated in place against the vapor pressure of pure water. The pressure was measured by means of calibrated precision pressure gauges (3-D Instruments). The pure components, water and ammonia, were charged to the cell with positive displacement pumps accurate to ±0.01 cc. Dissolved air and inerts were removed from the water and ammonia in the pumps by repeated evacuations of the cooled pumps. During each charge the temperature of the pumps was recorded in order to apply a density correction to the amounts charged.

The complete composition range was covered by dividing the procedure into two parts. First the cell was charged with pure water and the vapor pressure at the given temperature was measured. Weighed vapor samples were taken from the cell to remove any volatile impurities until the measured vapor pressure was constant. Once the vapor pressure of water was measured, ammonia was added incrementally in 5 mole percent steps. The system was degassed after each addition by drawing a very small vapor sample into a weighed cylinder. (The effect of this withdrawal on the equilibrium in the cell was accounted for in the data reduction procedure.) Once the system stabilized at the desired temperature, the pressure was read. This procedure was repeated up to a mixture composition of 55 mole percent ammonia. Then the contents of the cell were removed and the second part of the procedure was performed.

The second part of the procedure consisted of charging the cell with pure ammonia and adding 5 mole percent increments of water; each time performing the measurements as in the first part. This was done at 394.3 K and at 405.9 K. However, at higher temperatures, i.e., above the critical temperature of ammonia, a slight modification of this was required. A small amount of pure water was charged to the cell and data was taken by adding ammonia to reach high ammonia concentrations analogous to the first part of the procedure.

PTxy Measurements

PTxy data are obtained by analyzing samples of the two equilibrium phases. This sampling necessitates that the contents of the equilibrium cell be replenished so as to maintain the equilibrium concentrations initially present in the cell. This was accomplished using the flow apparatus shown in Figure 2.

The equilibrium unit was constructed of two 300 cc cylinders made of Type 316 stainless steel which were cast into an aluminum jacket and connected in series through a valve with stainless steel tubing. The aluminum jacket was thermally insulated. Two 1/16" stainless steel lines were welded into the top and bottom of each cylinder. One line from the bottom of each cylinder was used for charging or for sampling liquid. One line from the top of each cylinder was connected to a pressure gauge. One line connected the two cylinders from the top of the first to the bottom of the second and was held in contact with the aluminum jacket to maintain the same temperature as in the cylinders. One line from the top of the second cylinder was used for sampling vapor. One line in the bottom of the first cylinder was used for makeup of ammonia during vapor sampling. The temperature and pressure were measured as in the PTx measurements. The temperature probe was inserted in the center of the second cylinder. These cylinders were initially charged approximately half full with ammonia and water to the desired liquid composition.

The liquid in the second cylinder was sampled and analyzed by one of two methods depending on the composition. For high water concentrations the ammonia was absorbed into dilute sulfuric acid and the excess sulfuric acid was titrated. By this method, the amount of water in the sample was determined from the total sample weight less the weight of ammonia found by titration. For low water concentrations, the water was absorbed on Ascarite drying tubes and weighed, and the ammonia was absorbed on cobalt chloride hexahydrate followed by a Drierite drying tube to collect the water displaced from the cobalt chloride hexahydrate by the ammonia. These two tubes were then weighed together to determine the amount of ammonia in a given sample.

The vapor phase was sampled and analyzed with a sampling train consisting of two to six Ascarite drying tubes in series followed by a cobalt chloride hexahydrate tube and a Drierite tube or alternatively, the Ascarite tubes were followed by a flask of water and a Drierite tube. (The ammonia is absorbed into the water in the flask.) Drierite has an indicator for water that also acts as an indicator of ammonia, thus providing a method to detect any loss of ammonia from the end of the sampling train.

The pressure in the equilibration cells was maintained by charging ammonia to the bottom of the first cylinder. The ammonia bubbled up through the liquid in the first cylinder which served as a presaturator. The vapor from the first cylinder flowed into the second cylinder and bubbled up through the liquid there. The vapor from the top of the second cylinder was then sampled as an equilibrium phase. Liquid and vapor samples were drawn alternately until a repeatability of 5% in the relative volatilities was obtained.

MEASURED RESULTS

The total pressure (PTx) results at the eight temperatures studied are given in Tables 1 through 8 along with the derived VLE data obtained from the PTx data as described in the next section. Each table gives calculated phase compositions, measured and calculated pressures, activity coefficients for both ammonia and water, and the relative volatility of ammonia to water. Also shown in these tables are the Redlich-Kister activity coefficient equation parameters and the virial equation coefficients determined and used in the data reduction procedure. In these equations water is component 1 and ammonia is component 2.

Tables 9 through 16 present the results of the VLE measurements determined by the PTxy method at the eight temperatures studied. The composition results in these tables are averages of two to four analyzed samples. Given in these tables are the measured liquid and vapor-phase compositions, the calculated vapor-phase composition, measured and calculated pressure, calculated activity coefficients for both water and ammonia, the measured relative volatility of ammonia to water, and the relative volatility derived from PTx data.

DATA REDUCTION

Vapor composition data were derived from the total pressure PTx data by methods similar to those described previously (Gillespie et al. (11), and Barker (12)). In this procedure a correction was made for water and ammonia in the vapor phase in a material balance calculation based on the composition of the liquid phase. This required knowledge of the liquid-phase density. A correlation of the volume of the liquid as a function of composition was used which made a very small correction in the results. The data used to make this correlation are given in Table 17. The pure component data were taken from the literature and the mixture density data were taken from Perry's Handbook (2) at 288.2 K, but were measured as part of this project at the other temperatures. Other pure component data used in the data reduction procedure are given in Table 18. Table 19 compares measured pure component vapor pressures with literature values.

The Redlich-Kister activity coefficient equation with three parameters was used to calculate the activity coefficients of the system. Unsuccessful attempts were made to use the modified Wilson equation, the NRTL equation, and the new hyperbolic equation of Edwards et al. and Won et al. to fit the total pressure data. The Redlich-Kister equation was therefore used because of the flexibility it allows in fitting activity coefficient data.

An equation of state combining the hard-sphere repulsion term of the Redlich-Kwong equation and the 2nd and 3rd virial equation terms was used to calculate the vapor-phase fugacity coefficients of ammonia and water. The second virial coefficients for ammonia and for water were obtained from data tabulated by Dymond and Smith (15). One VLE data point was used to adjust the value of the second cross virial coefficient. Equations used in the data reduction procedure are outlined in the appendix.

DISCUSSION

Figure 3 shows measured total pressures as a function of liquid composition for each of the eight temperatures studied. As is seen in this figure, the ammonia-water system exhibits negative deviation from Raoult's law as is indicated by a negative deviation from a straight line through the end points on each isotherm. This deviation is most pronounced at low ammonia concentrations.

The activity coefficients of ammonia are given in Figure 4 as a function of liquid composition. Figure 5 similarly gives the activity coefficients of water. At all but the highest two temperatures, the activity coefficient of water is seen to have a minimum. This phenomenon is most likely real since it was required in order to obtain a good fit of the total pressure data and is the reason the Redlich-Kister equation was required to accurately model the data. Note the smoothness of the data and the consistent trend with respect to temperature. The infinite dilution activity coefficients for both ammonia and water are shown in Figure 6.

Relative volatilities are plotted as a function of liquid composition in Figure 7. The closed symbols are measured values and the open symbols are values derived from PTx data. A comparison of computed and measured values for all eight temperatures studied shows an average deviation of 6.0%.

With both PTx and PTxy data it is possible to verify the thermodynamic consistency of the PTxy measurements using the PTx measurements. Vapor compositions can be derived from the PTx data by implicit integration of the Gibbs-Duhem equation using the method of Barker (12) provided the vapor-phase nonidealities are correctly accounted for. This was accomplished by using one PTxy datum to adjust the second cross virial coefficient. In other words, one vapor point was required to accurately model the vapor phase. Once this was done a comparison of the data demonstrated the thermodynamic consistency of the PTxy results over the entire composition range. As seen in Figure 7, there is good agreement at all temperatures between measured relative volatilities and those derived from PTx data.

COMPARISON WITH LITERATURE DATA

Our measured Henry's Law constants for ammonia are compared to the correlation of Edwards et al. in Figure 8. These results are tabulated in Table 20. Good agreement is demonstrated at low temperatures, but at high temperatures there is significant deviation between their correlation and the experimental results. This difference at high temperatures is due to a paucity of data in this region at the time their correlation was developed.

Measured VLE data have been compared with literature data using relative volatility plots as shown in Figures 9 through 14. Figure 9 compares the correlated results of this study with the experimental measurements of T. A. Wilson along three isotherms. There is seen to be general agreement with these data. The average deviation for the points shown is 5.72%. It is important to note that although the results of our study agree well with the experimental measurements of T. A. Wilson, there are serious discrepancies between our results and those given in the Chemical Engineers' Handbook. The values listed in the handbook were obtained by smoothing and extrapolating T. A. Wilson's results. In terms of relative volatility, handbook values that are high by more than a factor of two are common. This discrepancy is probably a result of errors in the procedure used to smooth and extrapolate the data of T. A. Wilson.

The results of Macriss et al. at several temperatures are shown in Figure 10 where good agreement is again seen with deviations averaging 11.6%. Figure 11 compares data of Wucherer to the data of this project. The results of Wucherer agree fairly well with the data of this work, exhibiting an average deviation of 15.3%.

Figure 12 compares relative volatilities of Guillevic et al. to those of this work. Because the data of Guillevic et al. generally consisted of an analysis of only one phase at any given pressure, and since the composition of both phases is required to determine the relative volatility, the opposite phase composition was obtained by interpolating along the P-x diagrams given in their paper. To translate our results to the temperatures studied by Guillevic et al. the infinite dilution activity coefficients plotted in Figure 6 were used in conjunction with the correlation developed as part of this project. The average deviation between the results of Guillevic and those of this work is 14.1%. The agreement is good at low temperatures, but at higher temperatures the data of Guillevic et al. are consistently higher than our results.

Rizvi studied the ammonia-water system at sixteen temperatures. His results at three temperatures are compared in Figure 13 to the results of this work. Our data have been correlated to correspond to the temperatures studied by Rizvi. The average deviation for the points shown is 24.0%.

Figure 14 compares the isothermal data of Guillevic et al. and of Rizvi to the results of this work near the critical temperature of pure ammonia. Also shown in the figure is a curve representing the correlation developed as part of this research. At low concentrations of water the relative volatility must approach unity. This occurs because the vapor composition is identical to the liquid composition at the critical point. Since the temperature in question is the critical temperature of ammonia, the relative volatility becomes unity at pure ammonia.

The experimental data of this project agree well with the correlation. The data of Rizvi deviate significantly both above and below the curve, while the data of Guillevic et al. plot slightly higher than our data. However, the data of Guillevic et al. were obtained at a temperature about two degrees below the critical temperature of ammonia and therefore is expected to be somewhat higher than our results even though the general trend in the data as a function of concentration should be the same.

Based on the comparisons with literature data it appears that the results of this work are in good agreement with the older measurements of T. A. Wilson and Macriss et al. Fair agreement is shown with the data of Wucherer and the more recent data of Guillevic et al. and of Rizvi. A more in-depth comparison of the results of this study and the results of other researchers is warranted and would be of value to the chemical processing industry.

SUMMARY

PTx and PTxy data have been obtained for the ammonia-water system over the complete composition range at eight temperatures from 313.15 K to 588.7 K. The PTx data were reduced to PTxy data using the Redlich-Kister activity coefficient expansion with four parameters. A modified equation including a hard-sphere repulsion term was used to model the vapor phase. One PTxy datum was used to evaluate the second cross virial coefficient. The reduced data thus obtained was in excellent agreement with the actual PTxy measurements, giving a thermodynamically consistent set of vapor-liquid equilibrium measurements.

These results have been compared to literature data with varied results. Good agreement with the measurements of T. A. Wilson and of Macriss et al. is seen, but only fair agreement exists with the results of Wucherer, Guillevic et al., and Rizvi. Serious discrepancies exist with the extrapolated data from T. A. Wilson's measurements shown in Perry's Chemical Engineers' Handbook.

Literature Cited

1. Wilson, T. A., "The Total and Partial Pressures of Aqueous Ammonia Solutions, "Bulletin No. 146, Univ. of Illinois Engineering Experiment Station, Urbana, Ill. (1925).

2. Perry, R. H., and C. H. Chilton, eds., Chemical Engineers' Handbook, (1973).

3. Wucherer, J., "Measurements of Pressure, Temperature and Composition of Liquids and Vaporous Phase of Ammonia-Water Mixtures at Saturation Pressure Point," Z. Gesamt. Kalte-Ind. 41, 21-29 (1934).

4. Scatchard, G., L. F. Epstein, J. Warburton, Jr., and P. J. Cody, "Thermodynamic Properties - Saturated Liquid and Vapor of Ammonia-Water Mixtures," Refrigeration Engineering, 53, 413 (May 1947).

5. Macriss, R. A., B. E. Eakin, R. T. Ellington, and J. Huebler, "Physical and Thermodynamic Properties of Ammonia-Water Mixtures," Research Bulletin No. 34, Institute of Gas Technology, Chicago, Ill. (1964).

6. Edwards, T. J., J. Newman, and J. M. Prausnitz, "Thermodynamics of Vapor-Liquid Equilibria for the Ammonia-Water System," Ind. Eng. Chem. Fundam., 17, 264 (1978).

7. Won, K. W., F. T. Selleck, and C. K. Walker, "Vapor-Liquid Equilibria for the Ammonia-Water System," II International Symposium on Phase Equilibria and Fluid Properties in the Chemical Industry, Berlin, March 17-21, 1980.

8. Guillevic, J. L., D. Richon, and H. Renon, "Vapor Liquid Equilibrium Data for the Binary System Water-Ammonia 403.1 K, 453.1 K, 503.1 K up to 7.0 MPa," In Press (1983).

9. Rizvi, S. S. H., Ammonia-Water Equilibria, Ph.D. Thesis, University of Calgary, Calgary, Alberta (1985).

10. Bogart, M. J. P., Ammonia Absorption Refrigeration in Industrial Processess, Gulf Publishing Company, Houston (1981).

11. Gillespie, P. C., J. R. Cunningham, and G. M. Wilson, "Total Pressure and Infinite Dilution Vapor-Liquid Equilibrium Measurements for the Ethylene Oxide/Water System," Final Report AIChE DIPPR Project 805A (1981).

12. Barker, J. A., Determination of Activity Coefficients from Total-Pressure Measurements, Austrialian J. Chem., 6, 207 (1953).

13. Dymond, J. H., and E. B. Smith, The Virial Coefficients of Gases, Clarendon Press, Oxford, England (1969).

14. Orbey, H., and J. H. Vera, "Correlation for the Third Virial Coefficient Using Tc, Pc, and as Parameters," AIChE J., 29, 107 (1983).

15. Keenan, J. H., and F. G. Keyes, Thermodynamic Properties of Steam, Wiley, New York (1936).

Table 1. Total Pressure (PTx) Measurements and Derived Vapor-Liquid Equilibrium Data on the Ammonia-Water System at 313.15 K.

Mole % Water		Pressure, kPa		Activity Coefficient		Relative Volatility
Liquid	Vapor	Meas.	Calc.	H_2O	NH_3	NH_3/H_2O
100.00	100.000	7.38	7.38	1.000	0.146	26.26
98.43	70.007	10.43	10.38	1.000	0.150	26.86
97.44	58.222	12.36	12.37	0.999	0.152	27.31
94.71	38.279	18.23	18.32	0.997	0.161	28.87
89.88	21.300	31.77	31.21	0.986	0.183	32.81
84.85	12.609	49.14	49.32	0.964	0.214	38.82
79.81	7.729	73.46	74.31	0.928	0.255	47.19
75.04	4.946	103.66	106.40	0.882	0.304	57.78
69.88	3.093	147.63	152.83	0.821	0.367	72.68
64.89	1.993	206.40	211.64	0.753	0.438	90.91
60.12	1.329	278.44	282.29	0.684	0.515	111.94
55.72	0.929	381.59	360.59	0.619	0.590	134.17
55.10	0.885	335.79	372.63	0.610	0.601	137.48
50.73	0.633	490.92	464.13	0.548	0.678	161.73
50.20	0.608	476.43	475.98	0.541	0.688	164.74
45.76	0.442	612.81	580.68	0.483	0.764	189.97
45.18	0.425	599.74	594.99	0.476	0.773	193.22
40.76	0.316	739.47	707.40	0.424	0.842	216.88
35.70	0.231	864.91	839.61	0.375	0.910	239.88
30.64	0.172	983.56	969.28	0.335	0.962	256.08
25.58	0.130	1096.34	1090.27	0.307	0.997	263.31
20.51	0.099	1197.66	1198.46	0.289	1.015	260.02
15.42	0.074	1294.93	1292.63	0.284	1.019	245.30
10.28	0.052	1387.14	1375.69	0.295	1.013	218.86
5.14	0.030	1476.31	1453.79	0.328	1.005	182.36
0.00	0.000	1565.47	1538.80	0.397	1.000	139.39

Redlich-Kister Activity Coefficient Equation Parameters

$A = -1.97833 \quad B = -0.50000 \quad C = 0.55436$

Virial Coefficients

$B_{11} = -981.13 \quad B_{12} = -1078.48 \quad B_{22} = -211.65$, cc/mole

Table 2. Total Pressure (PTx) Measurements and Derived Vapor-Liquid Equilibrium Data on the Ammonia-Water System at 333.15 K.

Mole % Water		Pressure, kPa		Activity Coefficient		Relative Volatility
Liquid	Vapor	Meas.	Calc.	H_2O	NH_3	NH_3/H_2O
100.00	100.000	19.58	19.58	1.000	0.196	21.28
98.43	74.152	26.34	26.03	1.000	0.202	21.85
97.09	59.813	32.53	31.89	0.999	0.208	22.42
94.95	44.492	43.06	42.02	0.997	0.218	23.46
89.84	24.880	72.04	71.29	0.986	0.248	26.70
85.23	15.850	106.39	105.90	0.968	0.283	30.64
80.39	10.237	153.41	153.44	0.938	0.329	35.95
75.51	6.741	215.72	216.05	0.898	0.383	42.66
70.53	4.483	299.21	298.29	0.848	0.448	50.99
66.00	3.142	394.36	391.57	0.796	0.513	59.84
61.40	2.225	513.62	505.83	0.739	0.584	69.89
59.00	1.872	578.77	573.45	0.709	0.622	75.45
56.78	1.602	655.17	640.83	0.681	0.658	80.71
53.84	1.312	748.79	736.93	0.644	0.706	87.72
52.10	1.171	819.42	797.27	0.622	0.733	91.83
48.57	0.937	934.93	926.78	0.580	0.788	99.88
47.27	0.865	1000.79	976.58	0.565	0.807	102.68
43.16	0.681	1144.97	1139.80	0.520	0.864	110.73
37.65	0.506	1362.82	1366.57	0.467	0.928	118.76
32.45	0.390	1559.39	1579.77	0.428	0.974	122.59
27.01	0.302	1753.94	1791.93	0.397	1.005	122.07
21.52	0.234	1939.36	1987.07	0.380	1.020	116.67
16.04	0.179	2111.61	2160.19	0.379	1.021	106.56
10.70	0.129	2272.72	2312.19	0.397	1.014	92.65
5.27	0.074	2438.89	2462.14	0.444	1.005	75.28
0.00	0.000	2601.01	2623.48	0.533	1.000	56.69

Redlich-Kister Activity Coefficient Equation Parameters

A = -1.56693 B = -0.50000 C = 0.43797

Virial Coefficients

B_{11} = -730.86 B_{12} = -731.21 B_{22} = -177.66, cc/mole

Table 3. Total Pressure (PTx) Measurements and Derived Vapor-Liquid Equilibrium Data on the Ammonia-Water System at 353.15 K.

Mole % Water		Pressure, kPa		Activity Coefficient		Relative Volatility
Liquid	Vapor	Meas.	Calc.	H_2O	NH_3	NH_3/H_2O
100.00	100.000	47.35	47.35	1.000	0.264	17.29
98.95	84.254	55.73	55.70	1.000	0.269	17.61
96.67	61.202	75.28	75.22	0.999	0.282	18.40
94.68	48.119	95.35	94.02	0.997	0.295	19.19
89.99	29.575	146.41	146.38	0.987	0.331	21.41
85.22	19.230	211.26	214.09	0.970	0.375	24.22
80.56	13.087	291.41	297.86	0.945	0.425	27.52
76.22	9.348	386.15	394.58	0.916	0.477	31.08
71.61	6.662	507.13	519.86	0.878	0.538	35.34
67.02	4.841	655.37	669.91	0.834	0.602	39.94
62.24	3.537	847.68	854.32	0.786	0.672	44.95
62.13	3.513	810.09	858.91	0.785	0.674	45.06
57.67	2.669	1056.82	1057.30	0.738	0.739	49.68
57.13	2.585	1033.51	1082.93	0.732	0.747	50.22
52.61	1.997	1320.26	1310.10	0.684	0.810	54.48
52.04	1.936	1297.97	1340.24	0.679	0.818	54.98
47.81	1.549	1590.80	1572.69	0.637	0.872	58.22
46.46	1.448	1613.09	1649.72	0.624	0.888	59.07
41.25	1.131	1931.25	1955.41	0.578	0.942	61.38
35.81	0.894	2259.55	2280.10	0.539	0.985	61.86
29.94	0.707	2584.80	2622.65	0.508	1.015	60.02
24.02	0.564	2915.12	2947.08	0.491	1.028	55.79
18.18	0.446	3206.94	3240.43	0.493	1.027	49.59
12.18	0.332	3515.98	3519.03	0.520	1.018	41.61
6.10	0.200	3797.66	3798.69	0.586	1.006	32.40
3.59	0.131	3931.41	3922.22	0.630	1.002	28.44
1.15	0.047	4068.20	4052.05	0.685	1.000	24.58
0.00	0.000	4122.91	4117.89	0.717	1.000	22.78

Redlich-Kister Activity Coefficient Equation Parameters

A = -1.17411 B = -0.50000 C = 0.34079

Virial Coefficients

B_{11} = -565.88 B_{12} = -529.93 B_{22} = -151.23, cc/mole

Table 4. Total Pressure (PTx) Measurements and Derived Vapor-Liquid Equilibrium Data on the Ammonia-Water System at 394.25 K.

Mole % Water		Pressure, kPa		Activity Coefficient		Relative Volatility
Liquid	Vapor	Meas.	Calc.	H_2O	NH_3	NH_3/H_2O
99.08	89.892	225.35	227.19	1.000	0.442	12.11
97.24	73.746	275.20	273.32	0.999	0.460	12.54
95.58	62.546	330.42	318.46	0.998	0.477	12.95
90.95	41.514	482.41	463.87	0.991	0.528	14.16
86.40	29.160	658.41	638.03	0.978	0.582	15.43
81.50	20.715	884.57	864.47	0.960	0.643	16.86
77.21	15.756	1122.68	1098.53	0.940	0.698	18.11
72.79	12.144	1410.44	1376.26	0.915	0.755	19.35
68.02	9.375	1779.27	1718.01	0.887	0.814	20.56
63.16	7.366	2221.04	2109.64	0.856	0.872	21.56
61.01	6.666	2131.88	2296.05	0.842	0.895	21.91
58.06	5.853	2692.21	2563.81	0.823	0.926	22.27
56.20	5.413	2595.95	2739.22	0.811	0.944	22.42
52.92	4.750	3274.82	3059.53	0.791	0.972	22.54
51.29	4.467	3095.48	3223.30	0.782	0.985	22.52
47.56	3.912	3716.60	3607.43	0.762	1.010	22.28
46.33	3.753	3638.58	3736.39	0.756	1.018	22.14
41.16	3.193	4188.78	4286.49	0.734	1.041	21.21
35.84	2.755	4779.50	4857.96	0.720	1.054	19.72
30.25	2.397	5447.23	5454.35	0.716	1.057	17.66
24.14	2.081	6113.95	6095.45	0.730	1.050	14.97
18.56	1.825	6725.95	6677.05	0.764	1.037	12.26
12.59	1.535	7419.02	7318.72	0.832	1.021	9.24
6.51	1.115	8225.56	8044.66	0.954	1.007	6.18
4.00	0.825	8643.02	8383.49	1.026	1.003	5.01
1.28	0.326	9197.27	8782.95	1.123	1.000	3.96
0.00	0.000	9266.17	8981.48	1.177	1.000	3.59

Redlich-Kister Activity Coefficient Equation Parameters

$A = -0.52216 \quad B = -0.50000 \quad C = 0.18550$

Virial Coefficients

$B_{11} = -366.02 \quad B_{12} = -314.96 \quad B_{22} = -112.41$, cc/mole

Table 5. Total Pressure (PTx) Measurements and Derived Vapor-Liquid Equilibrium Data on the Ammonia-Water System at 405.95 K.

Mole % Water		Pressure, kPa		Activity Coefficient		Relative Volatility
Liquid	Vapor	Meas.	Calc.	H_2O	NH_3	NH_3/H_2O
99.02	90.326	310.05	323.12	1.000	0.492	10.82
96.59	71.396	396.18	402.01	0.999	0.519	11.35
95.53	64.850	430.63	439.35	0.998	0.532	11.58
90.88	44.058	620.11	625.88	0.990	0.588	12.65
86.14	31.084	841.00	857.24	0.978	0.647	13.78
81.86	23.370	1068.98	1104.85	0.963	0.702	14.80
77.15	17.539	1382.07	1421.94	0.943	0.763	15.87
72.41	13.470	1753.94	1789.00	0.919	0.822	16.86
67.73	10.614	2180.51	2197.73	0.894	0.877	17.68
62.87	8.469	2680.05	2667.86	0.867	0.929	18.30
59.83	7.434	2928.29	2983.43	0.851	0.958	18.55
57.84	6.855	3254.56	3198.05	0.840	0.975	18.64
55.11	6.167	3464.30	3501.74	0.826	0.997	18.68
52.68	5.642	3881.76	3780.11	0.814	1.014	18.62
50.37	5.207	4034.76	4050.61	0.804	1.028	18.48
47.34	4.716	4569.76	4412.57	0.792	1.043	18.17
45.53	4.459	4650.82	4631.86	0.785	1.051	17.91
40.55	3.866	5278.02	5243.11	0.772	1.064	16.96
35.46	3.394	5920.42	5873.08	0.766	1.069	15.64
30.16	3.004	6589.16	6529.79	0.771	1.066	13.94
24.69	2.674	7277.16	7207.88	0.792	1.056	11.93
19.06	2.374	7990.49	7918.13	0.835	1.041	9.68
13.23	2.067	8776.77	8700.07	0.911	1.023	7.22
6.90	1.615	9781.92	9675.88	1.050	1.008	4.52
4.32	1.258	10314.88	10133.61	1.130	1.003	3.54
1.74	0.599	11024.16	10621.04	1.228	1.001	2.94

Redlich-Kister Activity Coefficient Equation Parameters

A = -0.38154 B = -0.50000 C = 0.14968

Virial Coefficients

B_{11} = -328.92 B_{12} = -266.51 B_{22} = -104.01, cc/mole

Table 6. Total Pressure (PTx) Measurements and Derived Vapor-Liquid Equilibrium Data on the Ammonia-Water System at 449.85 K.

Mole % Water		Pressure, kPa		Activity Coefficient		Relative Volatility
Liquid	Vapor	Meas.	Calc.	H_2O	NH_3	NH_3/H_2O
100.00	100.000	910.10	910.10	1.000	0.615	6.91
99.03	93.530	955.80	967.52	1.000	0.631	7.06
97.56	84.577	1162.20	1060.86	0.999	0.655	7.29
96.02	76.219	1292.91	1166.94	0.998	0.681	7.53
91.55	56.973	1587.76	1523.93	0.991	0.752	8.18
87.46	44.428	1999.14	1915.34	0.982	0.816	8.72
83.03	34.661	2497.66	2408.64	0.969	0.880	9.22
78.59	27.641	3061.03	2972.73	0.955	0.938	9.61
74.17	22.557	3700.39	3599.12	0.939	0.990	9.86
74.02	22.410	3709.51	3621.44	0.939	0.992	9.86
71.35	20.035	4206.00	4029.87	0.929	1.019	9.94
69.41	18.556	4450.19	4339.25	0.922	1.037	9.96
67.79	17.457	4872.72	4605.14	0.917	1.050	9.95
64.55	15.573	5280.05	5156.08	0.906	1.074	9.87
63.18	14.885	5695.48	5396.15	0.902	1.083	9.81
59.40	13.263	6217.30	6078.27	0.892	1.103	9.57
57.74	12.661	6657.05	6386.05	0.888	1.109	9.43
53.97	11.497	7263.99	7101.42	0.881	1.120	9.03
50.64	10.671	7881.06	7750.77	0.877	1.126	8.59
48.42	10.207	8373.50	8191.52	0.877	1.127	8.26
41.97	9.194	9385.73	9508.52	0.882	1.122	7.14
31.88	8.593	11269.37	11717.60	0.922	1.094	4.98
21.15	0.000	13599.84	14735.24	0.000	0.000	1.00
11.67	0.000	17191.81	17810.23	0.000	0.000	1.00

Redlich-Kister Activity Coefficient Equation Parameters

A = -0.02460 B = -0.50000 C = 0.03927

Virial Coefficients

B_{11} = -231.51 B_{12} = -189.23 B_{22} = -79.34, cc/mole

Table 7. Total Pressure (PTx) Measurements and Derived Vapor-Liquid Equilibrium Data on the Ammonia-Water System at 519.26 K.

Mole % Water		Pressure, kPa		Activity Coefficient		Relative Volatility
Liquid	Vapor	Meas.	Calc.	H_2O	NH_3	NH_3/H_2O
100.00	100.000	3702.42	3702.36	1.000	1.050	5.22
98.95	94.772	3775.37	3900.84	1.000	1.054	5.20
97.94	90.179	4012.47	4094.89	1.000	1.058	5.18
96.99	86.207	4209.04	4279.98	1.000	1.061	5.16
94.09	75.827	4819.02	4859.43	1.000	1.069	5.08
90.80	66.542	5498.91	5540.24	0.999	1.074	4.96
86.81	57.832	6377.40	6394.36	0.999	1.077	4.80
83.50	52.141	7122.13	7122.99	0.999	1.076	4.64
79.35	46.433	8077.63	8059.76	1.000	1.072	4.43
74.68	41.398	9209.43	9143.50	1.002	1.065	4.18
69.96	37.388	10365.55	10273.17	1.005	1.056	3.90
64.02	33.446	11834.76	11755.70	1.011	1.043	3.54
57.86	30.295	13385.03	13391.48	1.020	1.029	3.16
51.00	27.559	15208.88	15381.60	1.031	1.016	2.74
43.70	24.935	17275.91	17762.46	1.045	1.004	2.34
36.54	21.388	19819.17	20402.82	1.056	0.996	2.12
34.24	19.975	20862.82	21328.62	1.060	0.995	2.09

Redlich-Kister Activity Coefficient Equation Parameters

$A = 0.09270 \quad B = 0.03779 \quad C = -0.08154$

Virial Coefficients

$B_{11} = -148.82 \quad B_{12} = -122.44 \quad B_{22} = -54.22$, cc/mole

Table 8. Total Pressure (PTx) Measurements and Derived Vapor-Liquid Equilibrium Data on the Ammonia-Water System at 588.75 K.

Mole % Water		Pressure, kPa		Activity Coefficient		Relative Volatility
Liquid	Vapor	Meas.	Calc.	H_2O	NH_3	NH_3/H_2O
100.00	100.000	10588.46	10588.25	1.000	1.138	3.29
99.22	97.518	10780.98	10883.21	1.000	1.131	3.24
98.00	93.943	11328.13	11343.00	1.000	1.121	3.16
96.59	90.224	11915.82	11871.55	1.000	1.109	3.07
92.50	81.466	13476.22	13385.15	1.002	1.073	2.81
87.69	74.027	15178.48	15119.84	1.007	1.032	2.50
82.33	68.262	17022.60	16989.14	1.014	0.989	2.17
76.35	63.894	19109.89	18982.82	1.026	0.947	1.82
72.79	61.416	20315.66	20101.96	1.034	0.926	1.68

Redlich-Kister Activity Coefficient Equation Parameters

A = -0.15310 B = 0.44466 C = -0.16227

Virial Coefficients

B_{11} = -104.30 B_{12} = -85.79 B_{22} = -38.35, cc/mole

Table 9. Vapor-Liquid Equilibrium Data (PTxy) on the Ammonia-Water System at 313.15 K.

Mole % Water		Pressure, kPa		Activity Coefficient		Relative Volatility
Liquid	Vapor	Meas.	Calc.	H_2O	NH_3	NH_3/H_2O
100.00	100.000	-	7.38	1.000	0.146	26.26
98.98	78.465	-	9.31	1.000	0.148	26.63
95.17	40.814	-	17.26	0.998	0.160	28.57
89.72	20.929	-	31.70	0.986	0.184	32.97
80.15	7.983	-	72.37	0.931	0.252	46.54
70.70	3.330	153.8	144.54	0.832	0.356	70.05
48.82	0.550	544.7	507.52	0.522	0.712	172.61
31.08	0.177	992.8	958.27	0.338	0.958	255.00
26.50	0.137	1103	1069.15	0.311	0.992	262.73
21.00	0.102	1227	1188.64	0.290	1.014	260.83
11.50	0.057	1406	1356.77	0.291	1.015	226.14
0.00	0.000	-	1538.80	0.397	1.000	139.39

Redlich-Kister Activity Coefficient Equation Parameters

A = -1.97833 B = -0.5000 C = 0.55436

Virial Coefficients

B_{11} = -981.13 B_{12} = -1078.48 B_{22} = -211.65, cc/mole

Table 10. Vapor-Liquid Equilibrium Data (PTxy) on the Ammonia-Water System at 333.15 K.

Mole % Water		Pressure, kPa		Activity Coefficient		Relative Volatility
Liquid	Vapor	Meas.	Calc.	H_2O	NH_3	NH_3/H_2O
100.00	100.000	-	19.58	1.000	0.196	21.28
98.99	81.930	-	23.68	1.000	0.200	21.64
95.25	46.254	-	40.54	0.998	0.216	23.30
89.78	24.727	-	71.69	0.986	0.248	26.74
80.06	9.945	-	157.18	0.936	0.332	36.36
70.55	4.491	305.4	297.92	0.848	0.447	50.95
48.90	0.956	958.4	914.30	0.583	0.783	99.15
30.35	0.353	1710	1663.48	0.415	0.988	122.95
23.62	0.259	1889	1914.99	0.385	1.016	119.31
19.67	0.215	2041	2047.90	0.378	1.022	113.76
9.37	0.116	2379	2348.70	0.406	1.012	88.65
0.00	0.000	-	2623.48	0.533	1.000	56.69

Redlich-Kister Activity Coefficient Equation Parameters

A = -1.56693 B = -0.5000 C = 0.43797

Virial Coefficients

B_{11} = -730.86 B_{12} = -731.21 B_{22} = -177.66, cc/mole

Table 11. Vapor-Liquid Equilibrium Data (PTxy) on the Ammonia-Water System at 353.15 K.

Mole % Water		Pressure, kPa		Activity Coefficient		Relative Volatility
Liquid	Vapor	Meas.	Calc.	H_2O	NH_3	NH_3/H_2O
100.00	100.000	-	47.35	1.000	0.264	17.29
99.02	85.173	-	55.14	1.000	0.269	17.59
95.15	50.808	-	89.42	0.997	0.292	18.99
89.89	29.294	-	147.64	0.987	0.332	21.46
79.94	12.459	-	310.51	0.941	0.432	28.00
69.02	5.552	575.7	601.32	0.854	0.574	37.90
48.93	1.641	1551	1509.75	0.647	0.858	57.44
28.59	0.671	2751	2698.85	0.502	1.019	59.25
25.64	0.600	2875	2860.92	0.494	1.026	57.16
23.26	0.547	2965	2986.77	0.490	1.028	55.09
18.69	0.456	3227	3215.85	0.492	1.028	50.20
17.54	0.434	3254	3271.04	0.494	1.026	48.81
10.44	0.297	3640	3597.93	0.534	1.014	39.07
7.07	0.224	3813	3752.87	0.572	1.008	33.92
0.00	0.000	-	4117.89	0.717	1.000	22.78

Redlich-Kister Activity Coefficient Equation Parameters

A = -1.17411 B = -0.5000 C = 0.34079

Virial Coefficients

B_{11} = -565.88 B_{12} = -529.93 B_{22} = -151.23, cc/mole

Table 12. Vapor-Liquid Equilibrium Data (PTxy) on the Ammonia-Water System at 394.25 K.

Mole % Water		Pressure, kPa		Activity Coefficient		Relative Volatility
Liquid	Vapor	Meas.	Calc.	H_2O	NH_3	NH_3/H_2O
95.61	62.726	310.26	317.61	0.998	0.477	12.94
91.51	43.487	434.38	444.65	0.992	0.522	14.01
74.52	13.414	1172.33	1263.08	0.925	0.732	18.88
49.30	4.156	3481.53	3426.81	0.771	0.999	22.43
31.70	2.482	5449.26	5300.37	0.716	1.057	18.24
8.29	1.265	-	7820.46	0.911	1.010	7.05

Redlich-Kister Activity Coefficient Equation Parameters

A = -0.52216 B = -0.5000 C = 0.18550

Virial Coefficients

B_{11} = -366.02 B_{12} = -314.96 B_{22} = -112.41, cc/mole

Table 13. Vapor-Liquid Equilibrium Data (PTxy) on the Ammonia-Water System at 405.95 K.

Mole % Water		Pressure, kPa		Activity Coefficient		Relative Volatility
Liquid	Vapor	Meas.	Calc.	H_2O	NH_3	NH_3/H_2O
95.47	64.504	427.49	441.52	0.998	0.532	11.60
90.33	42.223	630.85	650.52	0.989	0.594	12.78
73.80	14.519	1482.38	1676.40	0.926	0.805	16.58
49.20	5.007	4206.00	4189.51	0.799	1.034	18.37
31.28	3.080	6515.20	6391.19	0.769	1.068	14.32
8.01	1.719	9617.77	9490.89	1.020	1.010	4.98

Redlich-Kister Activity Coefficient Equation Parameters

A = -0.38154 B = -0.5000 C = 0.14968

Virial Coefficients

B_{11} = -328.92 B_{12} = -266.51 B_{22} = -104.01, cc/mole

Table 14. Vapor-Liquid Equilibrium Data (PTxy) on the Ammonia-Water System at 449.85 K.

Mole % Water		Pressure, kPa		Activity Coefficient		Relative Volatility
Liquid	Vapor	Meas.	Calc.	H_2O	NH_3	NH_3/H_2O
95.32	72.739	1254.40	1218.00	0.997	0.692	7.63
90.62	53.755	1634.37	1607.48	0.990	0.767	8.31
73.70	22.102	3481.53	3669.29	0.937	0.995	9.88
50.66	10.675	7660.17	7746.83	0.878	1.126	8.59
33.09	8.580	11183.24	11438.51	0.915	1.098	5.27
10.80	0.000	16688.23	18641.14	0.000	0.000	1.00

Redlich-Kister Activity Coefficient Equation Parameters

A = -0.02460 B = -0.5000 C = 0.03927

Virial Coefficients

B_{11} = -231.51 B_{12} = -189.23 B_{22} = -79.34, cc/mole

Table 15. Vapor-Liquid Equilibrium Data (PTxy) on the Ammonia-Water System at 519.26 K.

Mole % Water		Pressure, kPa		Activity Coefficient		Relative Volatility
Liquid	Vapor	Meas.	Calc.	H_2O	NH_3	NH_3/H_2O
95.40	80.219	4549.49	4595.09	1.000	1.066	5.11
90.08	64.790	5451.28	5692.19	0.999	1.075	4.93
73.63	40.424	9656.27	9391.57	1.002	1.063	4.12
52.10	27.960	-	15048.09	1.029	1.018	2.80
35.30	20.634	21784.88	20895.10	1.058	0.995	2.10

Redlich-Kister Activity Coefficient Equation Parameters

A = 0.09270 B = 0.03779 C = -0.08154

Virial Coefficients

B_{11} = -148.82 B_{12} = -122.44 B_{22} = -54.22, cc/mole

Table 16. Vapor-Liquid Equilibrium Data (PTxy) on the Ammonia-Water System at 588.75 K.

Mole % Water		Pressure, kPa		Activity Coefficient		Relative Volatility
Liquid	Vapor	Meas.	Calc.	H_2O	NH_3	NH_3/H_2O
96.76	90.651	12270.46	11808.01	1.000	1.110	3.08
90.62	78.249	13648.48	14069.54	1.004	1.057	2.69
74.10	62.374	20680.43	19696.96	1.031	0.933	1.73
68.30	57.621	20477.78	21420.74	1.044	0.903	1.58

Redlich-Kister Activity Coefficient Equation Parameters

A = -0.15310 B = 0.44466 C = -0.16227

Virial Coefficients

B_{11} = -104.30 B_{12} = -85.79 B_{22} = -38.35, cc/mole

Table 17. Molar Volume of Ammonia-Water Mixtures

Temp. K	Mole % H_2O	Molar Vol., cc/mole		Partial Molar Vol., cc/mole	
		Meas.	Calc.[a]	H_2O	NH_3
273.2	100	18.02[b]	17.95	17.95	21.37
310.9	100	18.15[b]	18.26	18.26	21.68
366.5	100	18.71[b]	18.74	18.74	22.16
422.0	100	19.63[b]	19.53	19.53	22.95
477.6	100	20.97[b]	20.89	20.89	24.31
533.2	100	22.95[b]	23.11	23.11	26.53
588.7	100	26.54[b]	26.48	26.48	29.90
288.2	100	18.04[b]	18.08	18.08	21.50
288.2	79.07	19.25[c]	18.94	17.91	22.81
288.2	58.63	20.36[c]	20.13	17.28	24.17
288.2	38.65	21.87[c]	21.76	15.93	25.43
288.2	0.00	27.56[c]	26.71	10.16	26.71
313.2	100	18.17[b]	18.27	18.27	21.69
313.2	66.46	20.74	19.93	17.62	24.51
313.2	40.33	22.72	22.37	14.98	27.36
313.2	0.00	29.41	30.84	-2.66	30.84
333.2	100.00	18.33[b]	18.43	18.43	21.85
333.2	69.62	20.51	19.94	17.82	24.81
333.2	37.27	23.36	23.46	13.32	29.48
333.2	0.00	31.25[d]	34.14	-12.90	34.14
353.2	100.00	18.54[b]	18.61	18.61	22.03
353.2	78.38	19.94	19.60	18.31	24.27
353.2	32.25	24.93	25.21	10.22	32.34
353.2	0.00	33.72[b]	37.46	-23.13	37.46
394.3	100.00	19.13[b]	19.08	19.08	22.50
394.3	82.73	20.01	19.86	18.86	24.63
394.3	29.78	28.00	27.88	5.38	37.42
394.3	0.00	45.53[d]	44.38	-44.05	44.38
405.9	100.00	19.33[b]	19.25	19.25	22.67
405.9	81.04	21.15	20.14	18.96	25.19
405.9	29.73	28.30	28.53	4.34	38.77
405.9	0.00	72.47[d]	46.39	-49.96	46.39
449.8	100.00	20.25[b]	20.12	20.12	23.54
449.8	69.70	23.07	22.00	18.91	29.13
449.8	28.88	33.76	31.52	-0.15	44.38
519.3	100.00	22.39[b]	22.46	22.46	25.88
519.3	76.45	24.71	23.85	21.67	30.92
519.3	28.81	38.07	36.79	-5.27	53.81

(Continued from Table 17)

(a) Molar Volume

$$\overline{V} = V_1^\circ + aY + bY^2$$

Component 1 = H_2O; Component 2 = NH_3

$$V_1^\circ = 17.95 + 0.008774T - 2.834 \times 10^{-5}T^2 + 2.73 \times 10^{-7}DT^3$$

Where T is in °C.

$$a = 1.9$$

$$b = -0.13 + 0.00616T$$

$$Y = \exp(1.8x_2) - 1$$

x_2 = liquid mole fraction of NH_3

$$\overline{V}_1 = \overline{V} - x_2 V_* ; \qquad \overline{V}_2 = \overline{V} + x_1 V^*$$

$$V^* = 1.8[\exp(1.8x_2)]\{1.9 + 2b[\exp(1.8x_2)-1]\}$$

(b) From Themodynamic Properties of Steam by J. H. Keenan and F. G. Keyes, John Wiley & Sons, N. Y. (1936).

(c) Perry's Chemical Engineers Handbook, Fourth Edition (1963), Table 3-51.

(d) F. Din, Thermodynamic Functions of Gases Vol. 1, Butterworth, Washington, D. C. (1962).

Table 18. Ammonia-Water Critical Constants[a] and Density Data Used in PTx Data Reduction

Compound	Tc, K	Pc, kPa	Acentric Factor	Density at 294.26 K kg/m^3
Ammonia	405.6	11,298	0.250	608.6[b]
Water	647.3	22,048	0.344	998.0[c]

(a) Reid, R. C., J. M. Prausnitz, and T. K. Sherwood, The Properties of Gases and Liquids. 3rd Ed., McGraw Hill Book Company, Appendix A, Property Data Bank, p. 630.

(b) Matheson Gas Data Book, Matheson Gas Products, 1974.

(c) ASME, Steam Tables, New York, 1967.

Table 19. Ammonia-Water Pure Component Vapor Pressures, Measured Compared with Literature Data

Compound	Temp. K	Vapor Pressure, kPa Meas.	Lit.	Percent Diff.
NH_3	313.15	1565	1556[a]	0.6
	333.15	2601	2617[a]	0.6
	353.15	4123	4148[a]	0.6
	394.3	9266	9302[a]	0.4
	405.7	-	-	-
	449.8	-	-	-
	519.3	-	-	-
	588.7	-	-	-
H_2O	313.15	7.377	7.375[b]	0.03
	333.15	19.58	19.92[b]	1.7
	353.15	47.35	47.36[b]	0.02
	394.3	-	205.6[b]	-
	405.9	-	293.4[b]	-
	449.8	910.1	928.2[b]	2.0
	519.3	3702	3722[b]	0.5
	588.7	10590	10640[b]	0.5

(a) Matheson Gas Data Book, Matheson Gas Products, 1974.

(b) ASME, 'Steam Tables', New York, 1967.

Table 20. Infinite Dilution Henry's Constant of Ammonia in Water

Temp. K	Henry's Constant, kPa/mole fraction	
	Measured	Correl. of Edwards et al[a]
313.2	190	179
333.2	446	388
353.2	828	763
394.3	2690	2470
405.9	2970	3300
449.8	7290	8880
519.3	19400	33700
588.7	33800	111200

(a) T. J. Edwards, J. Newman, and J. M. Prausnitz, AIChE J. 21, 248 (1975).

Table 21. Experimental Accuracy and Precision

The accuracy of measurements made are given below:

Measurement	Accuracy
Pressure	0.3%
Temperature	0.03 K
Pump Volume	0.01 cc
Sample Weights	0.1%
Relative Volatilities	5.0%

Anhydrous ammonia used in this study was obtained from Matheson Gas Company, Inc. with 99.99 mole percent purity. Distilled water which was degassed before charging was also used.

APPENDIX

NH_3-H_2O VLE CORRELATION

Component 1 is Water; Component 2 is Ammonia.

I. Basic Equation

$$P\phi_i y_i = f_i^\circ x_i \gamma_i \delta_i$$

where

P = total pressure
ϕ_i = Fugacity coefficient
f_i° = pure component fugacity at zero pressure
x_i = liquid mole fraction
y_i = vapor mole fraction
γ_i = liquid activity coefficient
δ_i = Poynting effect from zero pressure to saturation pressure of mixture:

$$\delta_i = \exp(P\overline{V}_i / RT),$$

where

$$\overline{V}_i = \text{partial molar volume}$$

II. Correlating equations:

Vapor

$$Z = \frac{1}{1-b/\overline{V}} - \frac{b}{\overline{V}} - \frac{b^2}{\overline{V}^2} + \frac{B}{\overline{V}} + \frac{C}{\overline{V}^2}$$

This equation has the hard sphere repulsion analoqous to the Redlich-Kwong equation of state, but also includes the 2nd and 3rd virial coefficients.

Equation of State Mixing Rules

Redlich-Kwong b:

$$b = y_1 b_1 + y_2 b_2 ; \quad b_1 = 21.11, \quad b_2 = 15.00$$

Second Virial Coefficient

$$B = y_1^2 B_{11} + 2 y_1 y_2 B_{12} + y_2^2 B_{22} ;$$

$$B_{11} = 18.02 [1.898 - (2641.62/T) \exp (186210 / T^2)]$$

(see reference #15)

$$B_{22} = 0.926 \{ 26.35 - 27.93 [\exp(725/T) - 1] \}$$

(Data from reference #13)

The leading constant, 0.926, was required to accurately correlate VLE data near the critical point of NH_3 (but used in all cases).

$$B_{12} = 1/2 [B_{11}^* (V_{c2}/V_{c1}) + B_{22}^* (V_{c1}/V_{c2})]$$

Here B_{11}^* means B_{11} calculated from the equation for B_{11}, substituting A_{12}/T for $1/T$. Similarly for B_{22}^*.

where,

$$A_{12} = \begin{cases} 0.944 + 0.0138(1000/T)^2 & T < 405.9 \text{ K} \\ 1.015 & T \geq 405.9 \text{ K} \end{cases}$$

This mixing rule for B_{12} is equivalent to assuming the following when $A_{12} = 1$:

$$B/V_c = y_1 B_{11}/V_{c1} + y_2 B_{22}/V_{c2}$$

$$V_c = y_1 V_{c1} + y_2 V_{c2}$$

or

$$B = y_1^2 B_{11} + y_1 y_2 [(V_{c1} B_{22}/V_{c2}) + (V_{c2} B_{11}/V_{c1})] + y_2^2 B_{22}$$

Third Virial Coefficient

$$\left.\begin{aligned} C_{111} &= 2097\ (cc/mol)^2 \\ C_{222} &= 4178\ (cc/mol)^2 \end{aligned}\right\} \quad \text{From } Z_c = 1 + B_c/V_c + C_{iii}/V_c^2$$

This equation forces agreement in Z_c at the critical point.

$$3C_{112} = 2V_{c2} C_{111}/V_{c1} + (V_{c1}/V_{c2})^2 C_{222}$$

$$3C_{122} = 2V_{c1} C_{222}/V_{c2} + (V_{c2}/V_{c1})^2 C_{111}$$

C_{ijk} consistent with

$$C/V_c^2 = y_1 C_{111}/V_{c1}^2 + y_2 C_{222}/V_{c2}^2$$

$$V_c = y_1 V_{c1} + y_2 V_{c2}$$

$$V_c^2 = y_1^2 V_{c1}^2 + 2y_1 y_2 V_{c1} V_{c2} + y_2^2 V_{c2}^2$$

which gives

$$C = y_1^3 C_{111} + 2y_1^2 y_2 c_{111} (V_{c2}/V_{c1}) + y_1^2 y_2 C_{111} (V_{c1}/V_{c2})^2 + y_1 y_2^2 C_{111} (V_{c2}/V_{c1})^2 + 2y_1 y_2^2 C_{222} (V_{c1}/V_{c2})^2 + y_2^3 C_{222}$$

and

$$C = y_1^3 C_{111} + 3y_1^2 y_2 C_{112} + 3y_1 y_2^2 C_{122} + y_2^3 C_{222}$$

so that

$$3y_1^2 y_2 C_{112} = 2y_1^2 y_2 C_{111} (V_{c2}/V_{c1}) + y_1^2 y_2 C_{222} (V_{c1}/V_{c2})^2$$

$$3y_1 y_2^2 C_{122} = y_1 y_2^2 C_{111} (V_{c2}/V_{c1})^2 + 2y_1 y_1^2 C_{222} (V_{c1}/V_{c2})$$

Liquid

Redlich-Kister Activity Coefficient Equation; 3 parameters

$$\frac{\overline{G}^E}{RT} = x_1 x_2 [A + B(x1-x_2) + C(x1-x_2)^2]$$

$$\ln \gamma_1 = x_2^2[A + B(x1-x_2) + C(x1-x_2)^2] + 2x_1 x_2^2 [B + 2C(x1-x_2)]$$

$$\ln \gamma_2 = x_1^2[A + B(x1-x_2) + C(x1-x_2)^2] + 2x_1 x_2^2 [B + 2C(x1-x_2)]$$

where

$$A = -18.676 + 22.9345(1000/T) - 8.8293(1000/T)^2 + 1.02863(1000/T)^3$$

$$B = \begin{cases} -0.5 & T \leq 450\ K \\ 3.485 - 1.79(1000/T) & T > 450\ K \end{cases}$$

$$C = -0.445 + 0.098(1000/T)^2$$

Fugacities

Water Calculated from pure component vapor pressures based on ASME Steam Tables(1967).

$$f_1^{\circ} = P_1^{\circ}\phi_1^{\circ}/\delta_1^{\circ}$$

Where P_1° is the vapor pressure of water at the temperature of the system. Taken from the ASME Steam Tables. ϕ_1° is the fugacity coefficient at P_1°.

Equations to calculate partial molar volumes needed for the determination of δ_i° are given in Table 17.

Ammonia An analytical equation was derived for the pure component fugacity which extends beyond the critical point of ammonia at 405.6 K.

δ_1° is the Poynting correction for water from zero pressure to P_1°.

and f_2° is given by

$$\ln f_2^{\circ} = 10.7215 - 4.9319(1000/T) + 1.4992(1000/T)^2 - 0.236202(1000/T)^3$$

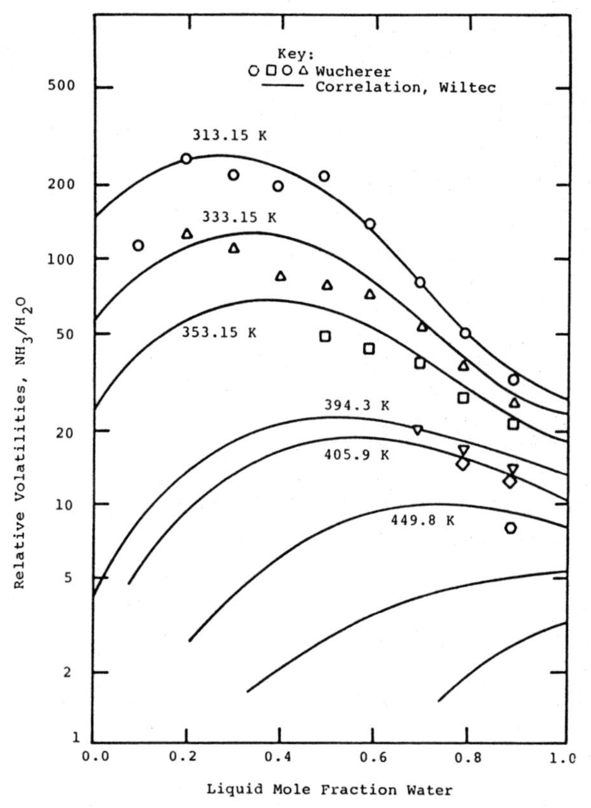

Figure 11. Relative volatilities as a function of liquid concentration. Data of Wucherer compared to new results for the ammonia-water system.

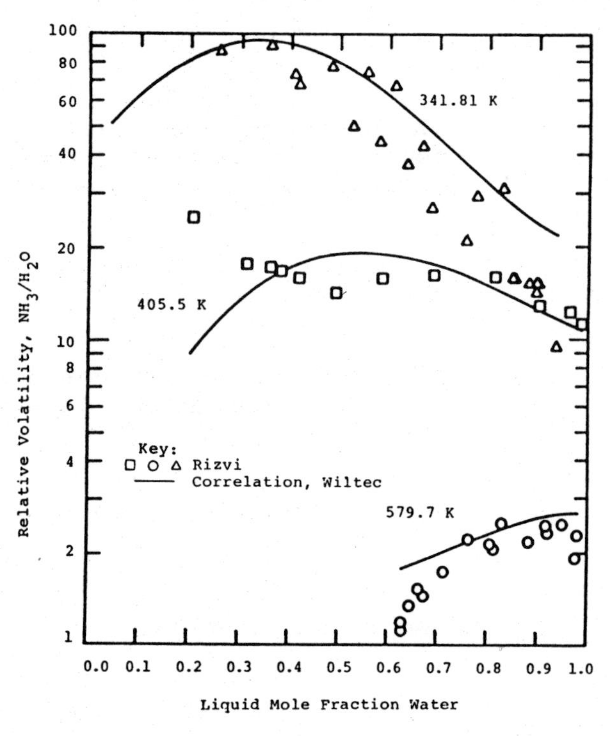

Figure 13. Relative volatilities as a function of liquid concentration. Data of Rizvi compared to new results for the ammonia-water system.

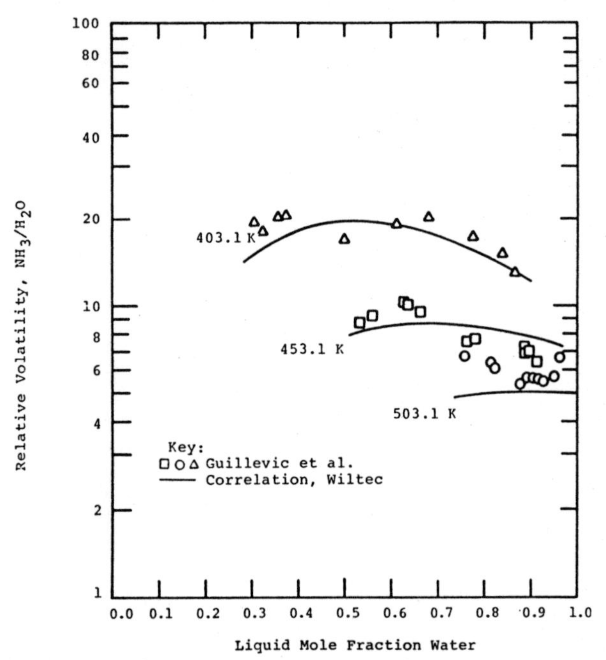

Figure 12. Relative volatilities as a function of liquid concentration. Data of Guillevic et al compared to new results for the ammonia-water system.

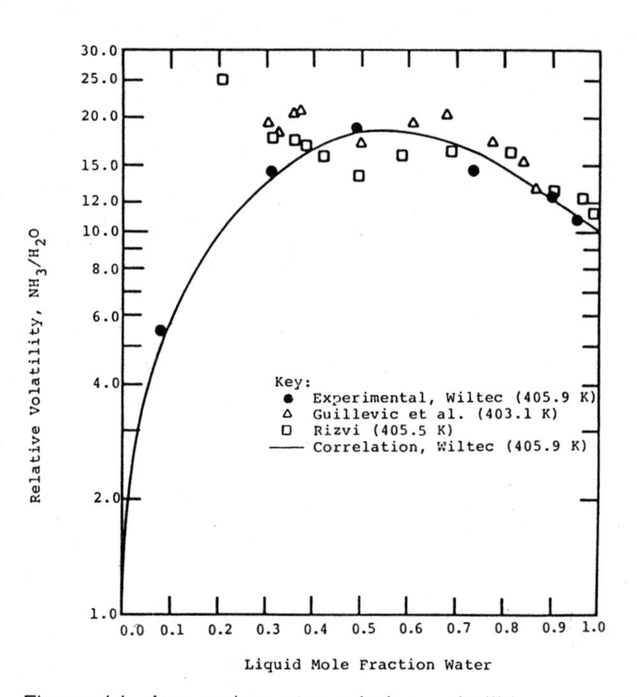

Figure 14. Ammonia-water relative volatilities near the critical temperature of ammonia.

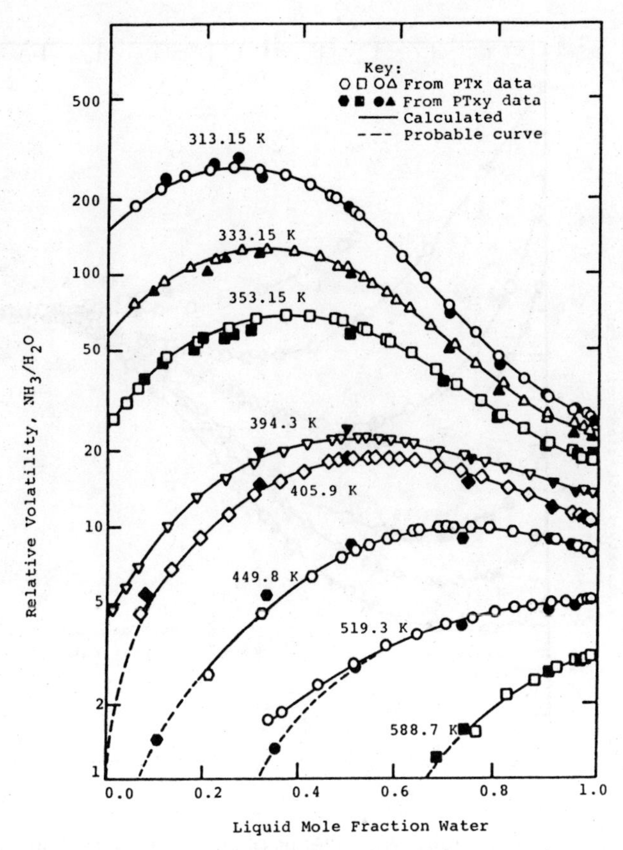

Figure 7. Relative volatility as a function of liquid concentration for the ammonia-water system.

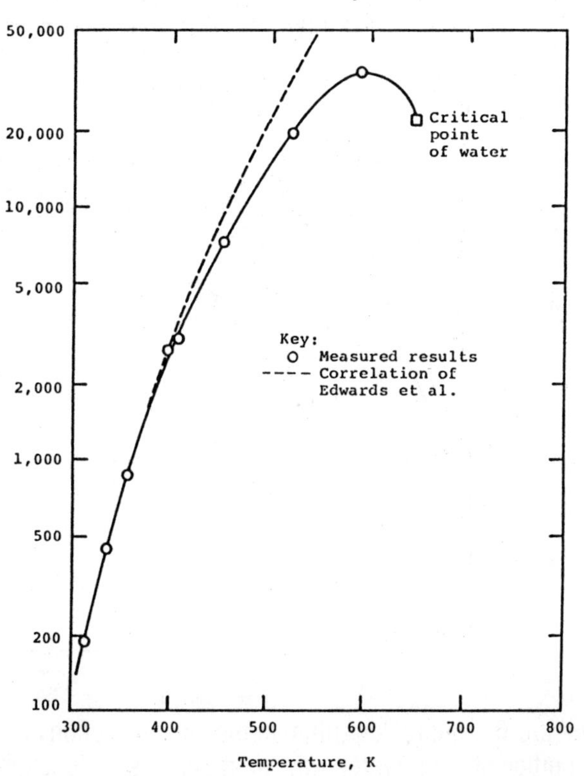

Figure 8. Henry's constants for ammonia in water. Measured results compared with correlation of Edwards et al.

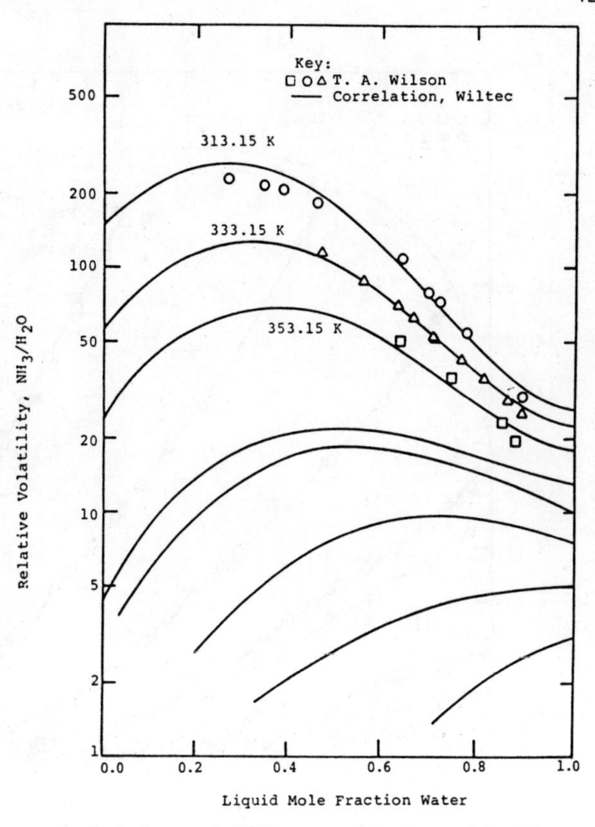

Figure 9. Relative volatilities as a function of liquid concentration. Data of T.A. Wilson compared to new results for the ammonia-water system.

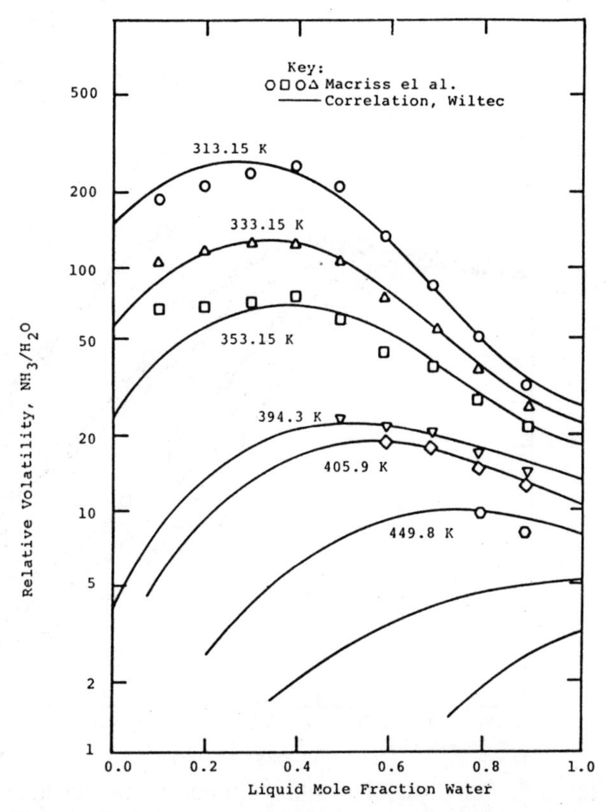

Figure 10. Relative volatilities as a function of liquid concentration. Data of Macriss et al compared to new results for the ammonia-water system.

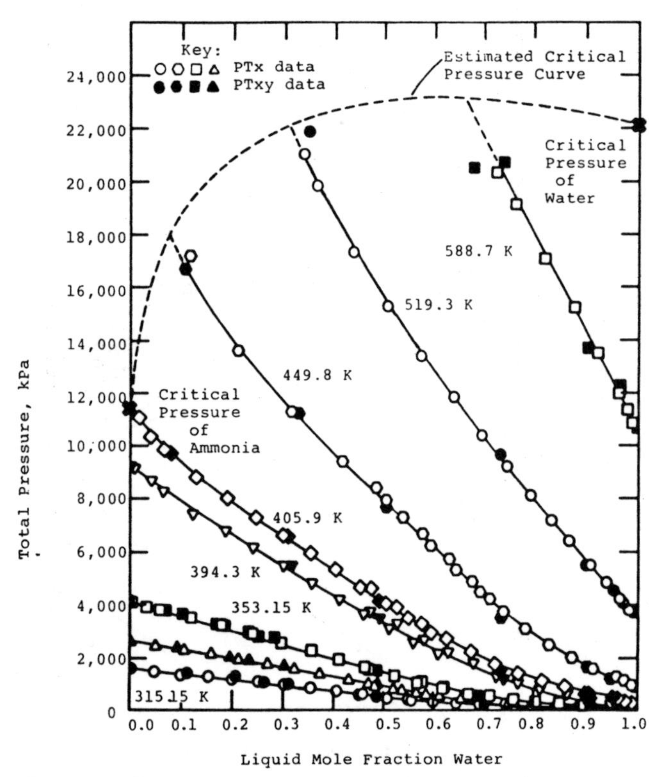

Figure 3. Total pressure as a function of liquid.

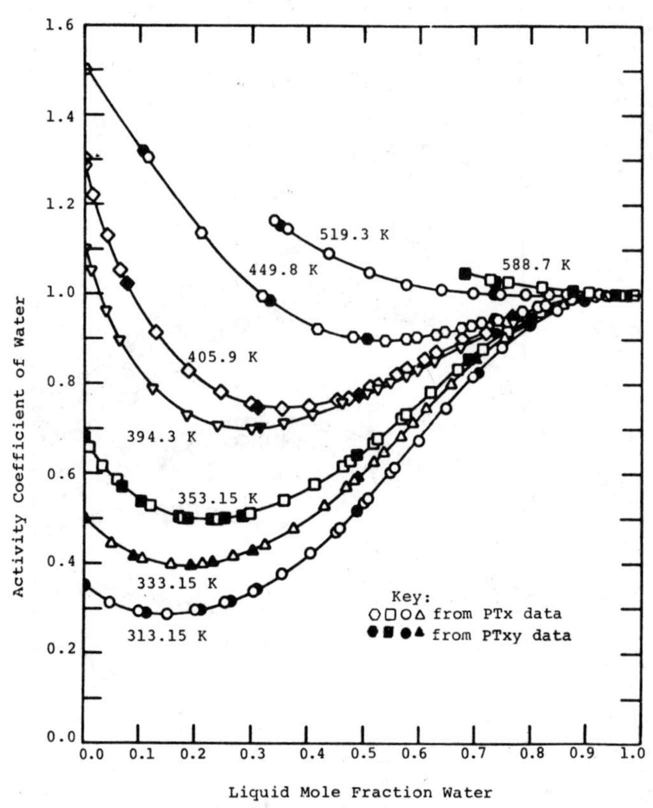

Figure 5. Activity coefficients of water as a function of liquid concentration.

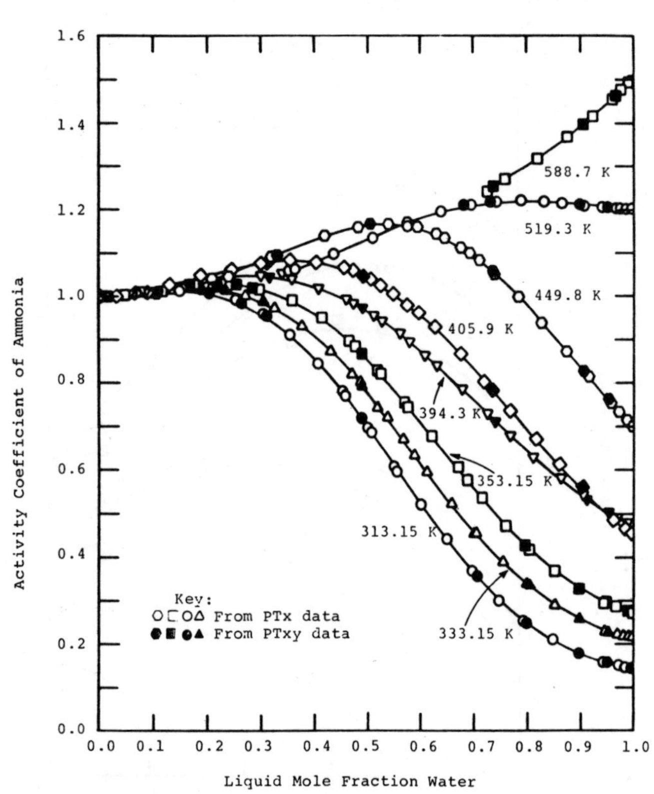

Figure 4. Activity coefficients of ammonia as a function of liquid concentration.

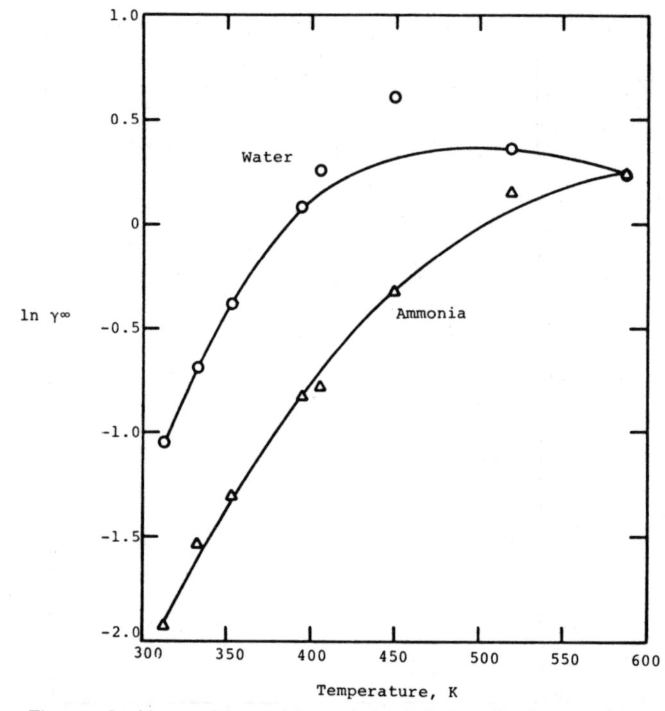

Figure 6. Natural logarithm of the infinite dilution activity coefficients for water and ammonia as a function of temperature.

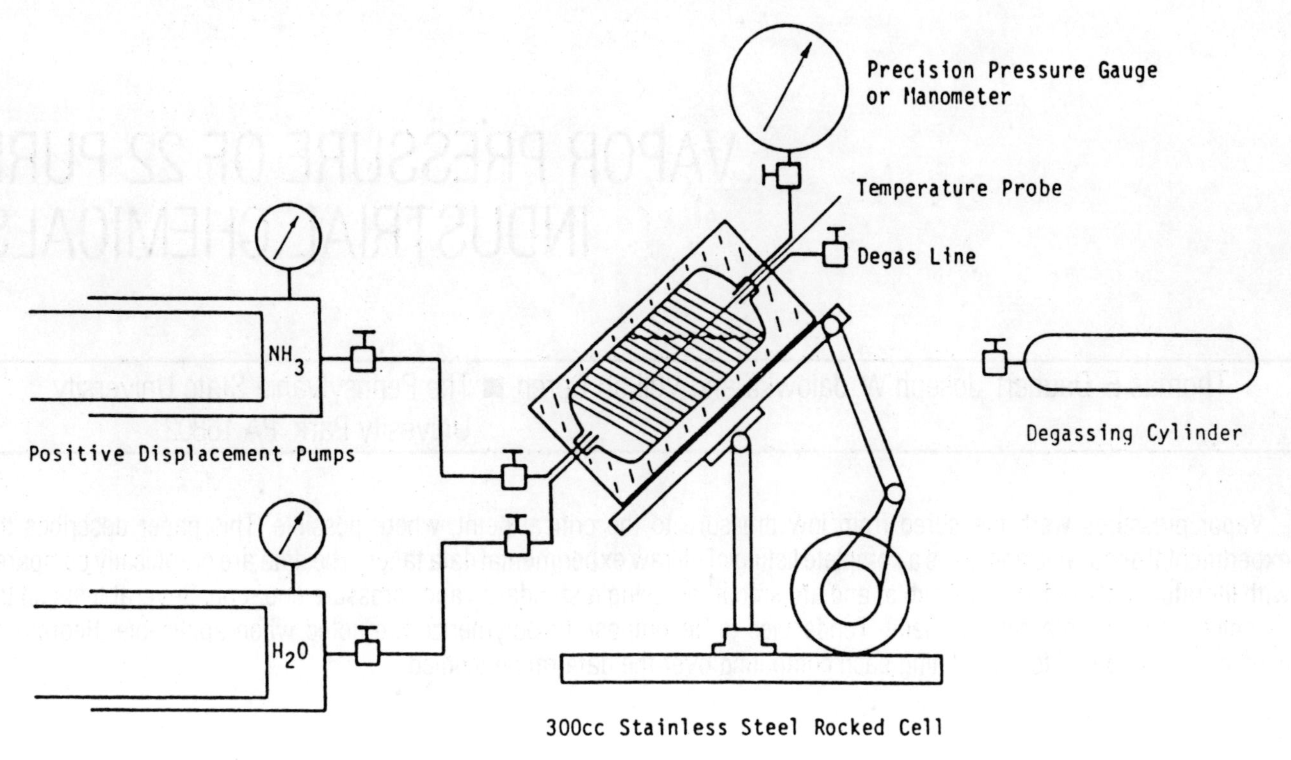

Figure 1. Schematic of rocked static cell apparatus for PTx measurements.

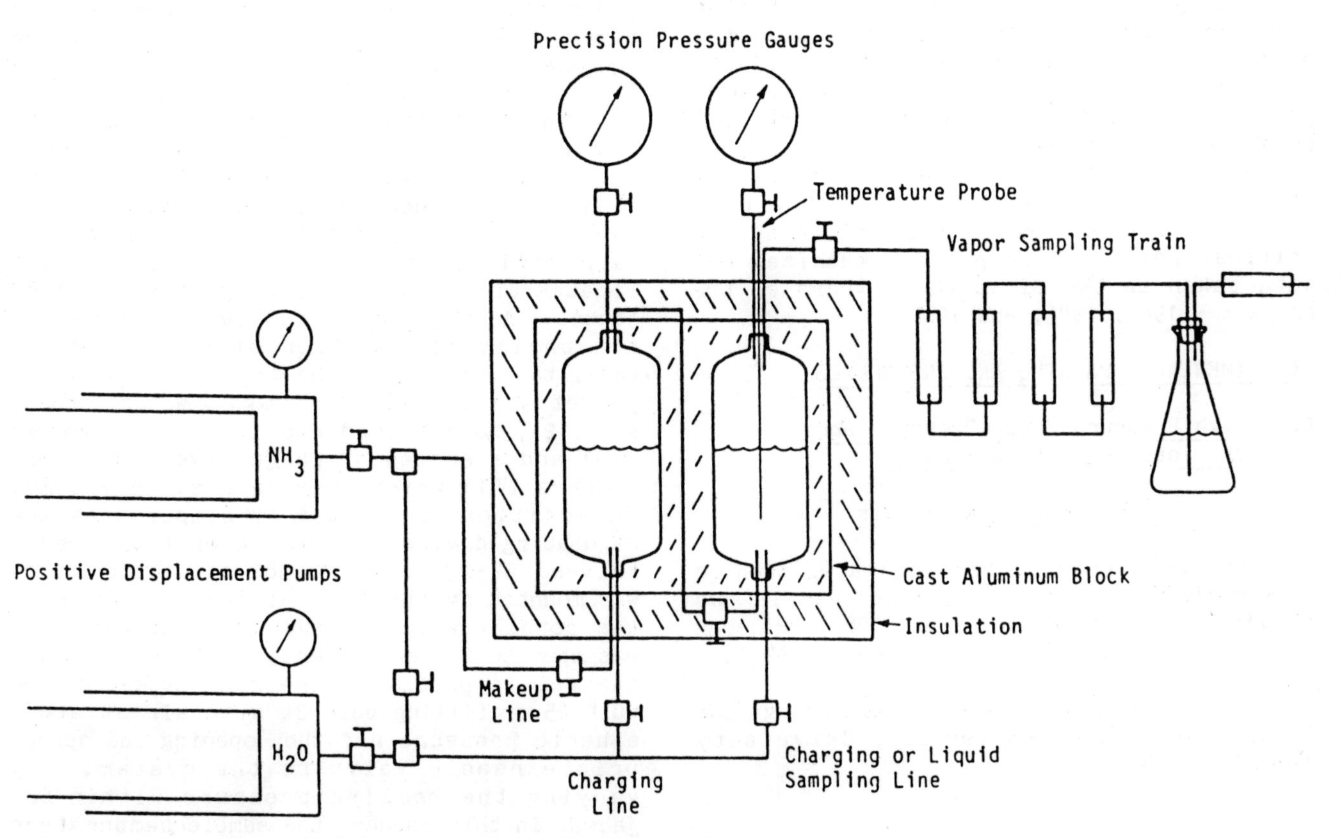

Figure 2. Schematic of flow cell apparatus for analyzing vapor and liquid samples.

VAPOR PRESSURE OF 22 PURE INDUSTRIAL CHEMICALS

Thomas E. Daubert, Joseph W. Jalowka and Volkan Goren ■ The Pennsylvania State University, Univesity Park, PA 16802

Vapor pressures were measured from low pressure to the critical point, where possible. This paper describes the experimental apparatus and gives a complete listing of all raw experimental data taken. The data are graphically compared with literature data (where applicable) and are smoothed using a standard vapor pressure equation. Several tests on the reliability of each data set are made. Tendencies to decompose or polymerize are noted when applicable. Regression coefficients are given for correlating each compound over the data range studied.

INTRODUCTION

Since 1982 the Design Institute for Physical Property Data has sponsored a project on the measurement of the liquid vapor pressure of pure compounds over a pressure range of 10 mm Hg to the critical point (where feasible). The work entails use of an ebulliometer for measurements below atmospheric pressure and a capillary tube apparatus from atmospheric pressure to the critical point. This paper summarizes the data taken in the first three years of the project - 1982, 1983, and 1984.

EXPERIMENTAL APPARATUS AND PROCEDURES

A. High Pressure Vapor Pressure and Critical Property Apparatus

1. Description of Apparatus:

The prototype of the apparatus used in this study was designed by Webster Kay in the 1930's (Kay, 1983), McMicking (1961), Hissong (1965, 1968), Pak (1968) and Spencer (1971).

Department of Chemical Engineering, The Pennsylvania State University, University Park, Pa. 16802.

The apparatus consists of three major sections: the pressure regulating and measuring section, the temperature control and monitoring section and the compression block assembly. A schematic diagram of the equipment is shown in Figure A.

The temperature of the sample was held constant by vapors of a series of pure organics which were boiled in the side-arm flask (4), connected to the bottom of the jacket. The temperature was measured by a calibrated copper constantan thermocouple (2) located in the upper portion of the heating jacket. An air stream (11) was directed at the unsilvered portion of the jacket to facilitate cooling. The top of the heating jacket was connected through a water condenser (12), to a closed-end mercury manometer (40) and a series of large five gallon containers (41) which provide a surge volume. In addition, a relatively simple pressure regulating device consisting of four needle valves (20-23), and two 80 milliliter bulbs was mounted on the control panel. One bulb was connected to a vacuum pump and the other was open to the atmosphere. Extremely small changes in pressure were made by evacuating bulb 25 or filling bulb 24 with air at atmospheric pressure and then opening the appropriate needle valve to the system. By varying the boiling pressure within the jacket in this manner, the sample temperature could be controlled very precisely. The mercury layer at the bottom of the heating jacket (9), was provided to prevent contamination of the boiling liquid by the rubber stopper (10).

The top of the thermocouple was placed about one inch above the closed end of the sample tube and the cold junction was kept in a Dewar flask containing an ice-water mixture. The thermocouple emf was measured with a Leeds and Northup type K-3 universal potentiometer.

The back leg of the u-shaped compressor block was connected by stainless steel tubing to the system inlet valve (33) located on the control panel. Pressure was transmitted to the sample by applying gas pressure to the mercury reservoir in the compressor block. Also, connected to the control panel were valves connected to a precision Heise Bourdon gauge (36), a two liter surge tank and a compressed nitrogen gas cylinder (18). The double valve system on the high pressure nitrogen inlet and exhaust lines of the control panel facilitated the precise control of the sample pressure.

2. Operating Procedure:

The preparation of sample and apparatus prior to obtaining data was careful and detailed.

Before starting an experiment all associated equipment was cleaned and assembled. Clean equipment, especially the sample tubes, was essential. The heating jacket was cleaned by scrubbing it thoroughly with a soap bubble, then rinsing it with cleaning solution. The cleaning solution was prepared by adding 15 grams of powdered potassium dichromate to 500 ml of concentrated sulfuric acid. The jacket was then rinsed successively with distilled water and acetone.

The sample tubes shown in Figure B were cleaned by allowing cleaning solution to stand in them for several hours. They were then rinsed with distilled water, acetone, and finally anhydrous ethyl ether. The tubes were placed in an oven at 325 K until they were used. The glass tubing used to make the capillary tubes was cleaned in the same manner as the sample tubes.

A small squeeze bottle, used in filling the sample tubes with mercury, was cleaned with nitric acid, rinsed with distilled water and acetone, and finally set aside to dry. The mercury was cleaned by distilling it twice. After distillation the mercury was stored in polyethylene bottles.

When starting the experiment, the sample tube was removed from the oven and allowed to cool. Two small rubber grommets were placed on either side of the expanded section.

The capillary tube, shown in Figure B, was flushed with the sample and then placed into the cool sample tube. Capillary tubes were made by drawing soft glass tubing to the desired length. The sample was introduced into the sample tube using the capillary tube as a funnel.

The amount of sample placed in the tube was determined by the expected experimental pressure range. Low pressure runs, 0-15 psig, required relatively little organic sample (height of fluid of 0.5-1.0 mm). If larger amounts of sample were used the vaporized sample had a greater volume than that of the visible portion of the sample tube and the dew point was not visible. High pressure experimental runs required a correspondingly greater amount of sample. At the critical point, a sample height of 10 mm was ideal. Problems obtaining the proper amount of organic sample in the sample tube were exacerbated by small amounts of sample between the mercury and the tube wall that rose and vaporized when the sample was heated.

The dew point rather than the bubble point was the criterion used to determine the vapor pressure of the sample. The bubble point was not used because if there was any soluble gas in the organic sample, it would escape from solution before the sample would vaporize.

After degassing the organic sample with the apparatus shown in Figure C the sample tube was filled with mercury using the procedure illustrated in Figure D. The sample tube containing the organic sample (A) was clamped in an inverted position over the mercury tray. A closed-ended capillary tube was inserted into the sample tube and mercury was introduced into the annular space between the capillary and sample tubes using a small squeeze bottle. The mercury fell only as far as the top of the capillary tube because of the capillary effect. The mercury was lowered in the sample tube (B). The capillary tube was removed resulting in (C), and an open-ended capillary tube was used to aspirate the excess organic from the mercury surface resulting in (D). The closed-ended capillary tube was then inserted into the sample tube to the depth of the mercury. The sample tube was then slowly filled with mercury. When the sample tube was being filled the capillary tube was rotated preventing air from being trapped between the mercury and the tube wall.

To facilitate inserting the sample tube in the front leg (6, Figure 1) a radiator hose was clamped on the end of the front leg. By increasing the pressure on the back leg (8, Figure A) to 4 psig, a pool of mercury formed in the radiator hose. The radiator hose increased the area of the front leg making it easier to insert the sample tube without allowing air to enter.

After tightly closing the end of the sample tube with the index finger, the tube was inverted and placed in the mercury pool. The pressure was released from the back leg and the sample tube was lowered in the front leg as the mercury level fell. The enlarged section of the tube then rested on a lip in the front leg approximately 10 centimeters below the top. Mercury was then aspirated from the front leg until it was equal to the level of the lip.

The gasket assembly was placed over the sample tube and held in place by a threaded collar. Spencer (1971) elucidated this construction fully. A rubber stopper was placed over the sample tube and seated on the threaded collar. The sample tube was then meticulously cleaned to prevent excess condensation from beading on it during the run.

The organic used for boiling was placed in the side arm boiler. Several glass beads were then inserted to prevent knocking. The heating jacket was then placed over the sample tube and seated on the rubber stopper. Approximately a centimeter of mercury was placed over the stopper at the base of the heating jacket. The mercury protected the stopper from the boiling organic during the run. The water condenser section of the heating jacket was attached and the thermocouple inserted. The water to the condenser (12, Figure A) and the air to the cooling air stream were then turned on. The vacuum line was attached and the heating assembly was evacuated and flushed with nitrogen or helium. The heating assembly was evacuated several times and filled with the inert gas until no oxygen was in the system. The potentiometer was calibrated and the barometric pressure was taken. The Heise gauge was standardized against the mercury manometer.

An electric heater (16, Figure A) was placed under the side arm boiler and supported by a scissors jack. At this point a safety shield was placed in front of the apparatus to protect the experimenter. For high temperature runs, 570 K and above, insulation was placed on the lower section of the heating jacket and a heating mantle was placed over the side arm boiler.

A pressure greater than the expected vapor pressure was charged to the system before heating. This prevented the organic from escaping the sample tube when heated. The electric heater was then turned on and the vacuum over the boiling organic was adjusted to produce the desired temperature.

3. Error Estimates

The temperature of the system was controlled by varying the pressure over the boiling organic. Small changes in temperature were effected by manipulation of valves (20, 21, 22) and (23). In Figure A using small inlet and outlet reservoirs (24 and 25) the vacuum over the boiling organic could be controlled to ±0.01 K. The apparatus was sufficiently leak-free that the temperature did not perceptibly change in a ten-minute period. This was important in order for the system to reach and maintain equilibrium.

The potentiometer could be read to ±0.005 mV which corresponded to ±0.01 K. The temperature of the copper block used in calibration varied on the order of ±0.5 K. Taking all of these effects into consideration it is estimated that the accuracy of the temperature values is ±0.1 K.

A Heise gauge was used for pressure measurements. The gauge was accurate to ±0.3 psi. The Heise gauge was calibrated using a Ruska dead weight tester. The pressure gauge corrections were obtained by linear interpolation between specific calibration points. Because of the additional error introduced by this interpolation, the accuracy of pressure values was estimated to be ±0.5 psi.

B. Low Pressure Vapor Pressure Apparatus

1. Description of Apparatus

To obtain the vapor pressure of various compounds at temperatures below their normal boiling points, the apparatus pictured in Figures E and F was constructed. This improved simple ebulliometer was first designed by W. Swietoslawski and W. Romer at the Polytechnic of Warsaw in 1942.

The sample is heated by a 200 watt heater which fits into a well in the bulb and an electrical insulator which is wrapped around the bulb. The vapor from the boiling sample travels up and directed to the well where an iron constantan thermocouple measures the temperatures. At this point some of the vapor condenses and recycles back into the bulb. The remaining vapor condenses in a vertical condenser. The condensate from the condenser falling back on to a drop counter is also recycled into the bulb. The purpose of the drop counter is to find the true boiling temperature of the sample at that pressure. Usually the temperature of the system increases as the number of drops increases and reaches a steady state temperature. At this point a large change in the heat input will result in no significant temperature change. Therefore, temperature and pressure measurements can be made at a consistent boiling point.

2. Operating Procedure

Before starting an experiment all the equipment is cleaned and assembled. Clean equipment is essential since impurities effect the vapor pressure of the sample. The ebulliometer is filled with nochromix added to sulfuric acid and let sit overnight. Then it is rinsed with distilled water and put into the oven to dry.

The sample is introduced into the ebulliometer and filled up to the top of the bulb with the heating element in the well. The glassware is then fitted together and sealed using vacuum grease. With the vacuum line the system is flushed with nitrogen gas to remove the air since some compounds might react with oxygen. Next the ebulliometer is wrapped with an insulation tape leaving the drop counter and part of the recycle system open. After adjusting the pressure system at the lowest pressure required, the sample is heated and boiled by turning on the heater and the insulation. The amount of heat is increased up to a steady temperature by monitoring the drop counter. Once even boiling and steady temperature are achieved the temperature and pressure are recorded. The temperature is read from a digital thermometer (Omega Eng. Inc.) in connection with an iron constantan thermocouple and the pressure is read from a mercury manometer.
The pressure is then increased and the procedure is repeated.

3. Error Estimates

The mercury level in the manometer is measured with a cathetometer which has an accuracy of 0.005 cm Hg. The temperature is corrected according to the thermocouple's individual calibration curve. The digital thermometer by Omega Eng. Inc. has an accuracy of 0.1°C.

C. Sources of Chemicals

Most chemicals were provided by the sponsors at the highest purity available.

	Compound/ Company	Yr.
1.	Isopropyl Alcohol Aldrich Chemical, Lot 1025HH	82
2.	Phenol Allied Chemical, Lot 3049	82
3.	*n*-Dodecylbenzene	84
4.	*p*-Ethyltoluene Aldrich Chemical, Lot 3402LH	82
5.	Methyl-*t*-butyl ether Eastman Kodak, Lot D11A	83
6.	Monoethanolamine Texaco Chemical, Lot 7H-0561/ H5-0267	82
7.	Diethanolamine Texaco Chemical, Lot 7H-2209	83
8.	2,4-Toluenediamine Olin Corp.	83
9.	Hexamethylenediamine Aldrich Chemical, Lot 1201PJ	83
10.	Hexamethyleneimine Aldrich Chemica, Lot 042064	83
11.	Diethylene glycol Olin Corp.	84
12.	Triethylene glycol Fluka Ag Chemical Co., Lot 22388 1280	82
13.	Propylene glycol monomethyl ether acetate Eastman Kodak, CAS #108-65-6	84
14.	Vinyl acetate Celanese Chemical	82

15. Dimethyl succinate 84
    Amoco Chemicals

16. Ethyl acrylate 83
    Celanese Chemical

17. Methacrylic acid 82
    Rohm and Haas, Lot B67372

18. ε-Caprolactam 84
    Allied Fibers

19. N-Cyclohexylpyrrolidone 82
    Olin Corp.

20. Dimethyl sulfoxide 84
    Aldrich Chemical 99.9% pure Lot 8324AL

21. Trimethoxysilane 84
    Dow Corning Corp.

22. 3,3,3-Trifluoropropene 82
    Dow Corning Corp.

MEASURED RESULTS

Table 1 summarizes the compounds studied giving the range of temperature and pressure, the measured critical conditions where applicable, and notes concerning certain experiments. Table 2 lists all raw experimental data in units of °C and pascals. Each experiment carried out is listed separately. Smoothing was carried out by fitting the data to an equation as discussed in the next section.

DATA ANALYSIS

The general procedure of data analysis primarily consists of (1) graphical comparison of the data taken with literature data where available, (2) fitting of the data with a regression equation which also smooths the data, and (3) determining if the signs of the coefficients of the regression equation show a correct shape of the logarithm of vapor pressure - reciprocal temperature curve. A plot for each compound is given in this paper and discussed together with data qualification.

Qualification of Experimental Data

For each compound four qualification tests generally accepted from past work (Chase, 1984) were used to analyze the experimental data.

(1) The Sign Rule - The experimental data when regressed to the DIPPR Data Compilation (Daubert and Danner, 1985) regression equation

$$P = \exp(A + B/T + C \ln T + DT^E)$$

must follow the convention that the A coefficient is positive, B and C are negative, and D and E are positive for thermodynamic consistency. For this purpose a constrained fitting program supplied by J. David Chase of Celanese was used. This program is constrained at the critical point. If experimental criticals were not available, predicted values were used. It should be noted that if the equation were regressed unconstrained, the sign rule may not be passed even though the data are accurate as for many of the compounds the sparcity of data above the decomposition point to near the critical could cause the shape of the curve to be predicted erroneously. Passing of the sign rule only signifies that the data may be accurate. Coefficients derived for each compound are given in Table 3.

(2) The Line Rule - Using a plot of $\ln P$ vs. $1/T$, the experimental data are compared with a line drawn from the melting point to the critical point. Both of these points are obtained from the experimental data, prediction, or from the DIPPR Compilation Project. Values used for criticals are given for each compound in Table 3 with the coefficient. The line rule states that the experimental data must coincide or deviate positively from the line connecting the melting point and the critical point.

(3) $\Delta H/\Delta Z$ Plot - Plotting $\Delta H/\Delta Z$ (equal to $RT^2 \, d \ln P/dT$) vs. reduced temperature a minimum must exist in the range of 0.8-0.9 reduced temperature as shown by Ambrose (1980).

(4) Family Plot - For compounds for which the experimental critical point is reached, plots of T_c and P_c vs. normal boiling point for the compound's homologous series is plotted where experimental data are available on other series members to indicate consistency of the series.

Space limitations dictate that only the plot showing the experimental data both measured and from the literature together with the regression line can be included. The line rule, $\Delta H/\Delta Z$, and family plots are not illustrated although comments are made whenever applicable.

ANALYSIS BY COMPOUND

1. Isopropyl alcohol - Isopropyl alcohol was chosen as the base compound for this work as both accurate vapor pressure and critical property data were available. The experimental data taken in this work compared favorably with the previous experimental data as shown in Figure 1. When the regression equation was fitted to the data, the coefficients shown in Table 3 were obtained passing the sign rule. The line rule using experimental end points shown in Table 3 was also passed. The minimum in the $\Delta H/\Delta Z$ plot was at a reduced temperature of 0.90. A family plot of the isoalcohols from C_3 to C_8 showed that the experimental critical temperature for isopropanol coincided with the series-qualified regressed critical temperature.

2. Phenol - Figure 2 indicates the excellent consistency of the experimental data taken in this work with the literature values as well as its extension to higher temperatures. Measured vapor pressures are slightly lower at the higher temperatures than those which were previously predicted. Use of the critical from DIPPR Compilation gives regression coefficients in Table 3 which pass the sign rule. However, when the line rule was implemented the data lie below the line indicating the predicted critical to be too high. A minimum in the $\Delta H/\Delta Z$ curve occurs at a reduced temperature of 0.86.

3. n-Dodecylbenzene - A plot of the data in Figure 3 shows consistency of the data to be excellent in the range above atmospheric pressure. Samples of pure straight chain material could not be obtained except in a very impure state to extend the data to lower pressures. The data taken passed the sign and line rules and had a minimum at a reduced temperature of 0.84 on the $\Delta H/\Delta Z$ plot.

4. p-Ethyltoluene - Except for minor scatter near the critical point, experimental data agreed with past sources of data over the entire range as shown in Figure 4. The critical point was not previously measured. The sign and line rules were passed. A minimum in $\Delta H/\Delta Z$ was obtained at a reduced temperature of 0.84. Series qualified critical temperature and pressure was determined using the C_7-C_{10} members of the homologous series. The experimental critical temperature and pressure are essentially identical to or just slightly higher than the series qualified counterparts indicating both consistent and accurate criticals.

5. Methyl-t-butyl ether - Consistency of the measured data within the data set and with previously reported data are excellent as shown in Figure 5. The measured criticals are essentially identical to those measured by Ambrose (1035). The regression coefficients passed the sign rule. The line rule test and minimum in the $\Delta H/\Delta Z$ plot at a reduced temperature of 0.84 are satisfactory. Using C_2-C_5 ethers, series qualification of both the critical temperature and the critical pressure was successful.

6. Monoethanolamine - Data are consistent within the data set. As shown in Figure 6, at temperatures near the decomposition point (~350°C), the observed vapor pressures are somewhat lower than those earlier predicted (Riedel, 1257). Data at lower temperatures are completely consistent with previous work. The sign rule was passed as was the line rule up to the point where decomposition occurred. The $\Delta H/\Delta Z$ plot yielded a minimum at a reduced temperature of 0.86.

7. Diethanolamine - Consistency of the measured data with low temperature experimental data taken previously are excellent as shown in Figure 7. Decomposition occurred above a temperature of 244°C which is considerably below the compound's normal boiling point. The sign rule was passed as was the line rule. A minimum at 0.84 reduced temperature occurred in the $\Delta H/\Delta Z$ plot.

To complete the ethanolamine series, attempts were made to measure the vapor pressure of triethanolamine. This compound decomposed violently at its normal boiling point of about 340°C and relatively rapidly below one atmosphere preventing acquisition of data.

8. 2,4-Toluenediamine - Two experimental references and the measured data showed very good consistency as shown in Figure 8. The sign rule and line rules were passed, and a minimum in the $\Delta H/\Delta Z$ plot was obtained at a reduced temperature of 0.85.

9. Hexamethylenediamine - Figure 9 shows a plot of the data taken in this work. The data are quite consistent up to the point where decomposition begins (about 291°C). No experimental data are given in the literature except a point of 30 Pa at 20°C of unknown reliability. However, the data are quite consistent with the Riedel (1257) predictor above one atmosphere. The data when regressed pass the sign rule. The line rule

is passed and a minimum in the ΔH/ΔZ curve occurs at a reduced temperature of 0.86.

As a part of this work hexamethylenetetramine vapor pressure measurements were attempted. The compound sublimes at about 200°C, below the melting point; thus, vapor pressure data is unobtainable.

10. Hexamethyleneimine - The measured data when plotted together with two references of experimental data in Figure 10 show excellent consistency up to the decomposition point of about 347°C. The regression coefficients pass the sign rule while the line rule check was satisfactory. A minimum in the ΔH/ΔZ curve at a reduced temperature of 0.82 exists.

11. Diethylene glycol - The experimental data together with three sets of other experimental data from the literature are plotted in Figure 11. Vapor pressure data are quite consistent below 3.8 atmospheres. At the higher pressures, experimental vapor pressures from this work are slightly lower than those from the cited reference (Rikenbach, 409). Polymerization appeared to begin for the highest temperature runs (9982). When the line rule using the predicted criticals was checked, the highest pressure data set failed as it lay below the line. As this data was suspected to be slightly low because of decomposition/polymerization only two of the three sets of data were used for regression. The regression coefficients passed the sign rule and the ΔH/ΔZ plot showed a minimum at a reduced temperature of 0.86.

12. Triethylene glycol - Data taken when compared with two other data references in Figure 12 are consistent at low pressures but somewhat low at the highest temperature set of data (9999). Decomposition/polymerization of the sample may have been occurring at these conditions and was definitely taking place at the highest temperature of 346°C. However, the sign rule was passed. The line rule failed using the predicted critical indicating either too high a value for the critical or that the vapor pressure data at high temperatures is too low. The ΔH/ΔZ plot shows a minimum at a reduced temperature of 0.86.

Both for DEG and TEG, the criticals predicted by the Lydersen method are probably not very accurate. Thus, the extent of the error in high temperature vapor pressure may not be as high as would be indicated by the plots.

13. Propylene glycol monomethyl ether acetate - As no experimental literature data exist for this compound, Figure 13 shows only the data taken in this work. Consistency of subatmospheric and superatmospheric data is very good. The highest temperature point appears to be high with decomposition beginning. Regression coefficients pass the sign rule. The plot of ΔH/ΔZ shows a minimum at a reduced temperature of 0.86.

14. Vinyl acetate - Vinyl acetate data were taken to the critical point and are plotted with five experimental data references in Figure 14 showing good consistency. The experimental criticals are 5°C lower and 0.06 MPa higher than the values predicted by Lydersen and are reproducible. The sign and line rules were passed. At a reduced temperature of 0.86 the ΔH/ΔZ plot showed a minimum. Series qualified criticals for both T_c and P_c using C_c-C_7 acetates both normal and isomeric shows both experimental values just slightly above the series qualified critical validating its consistency and accuracy.

15. Dimethyl succinate - As no experimental literature data are available Figure 15 only shows the four sets of experiments carried out in this work. Data consistency and fit to the regression equation are excellent. The sign and line rules were passed while a minimum in the ΔH/ΔZ plot occurred at a reduced temperature of 0.84.

16. Ethyl acrylate - The experimental data taken in this study are plotted together with data at low temperatures from three literature references in Figure 16. The low temperature data from this work are slightly lower than data taken from the literature but within any reasonable error band. The higher temperature data of this work are consistent within themselves and with low temperature data up to the point where polymerization begins to occur and the vapor pressure begins to fall (about 235°C). The sign and line rules are passed with the ΔH/ΔZ plot showing a minimum at a reduced temperature of 0.86.

17. Methacrylic acid - Experimental data on methacrylic acid were only taken in the capillary apparatus from 154-256°C. Attempts to take data below 160°C in the ebulliometer resulted in appreciable polymerization even when stabilized with hydroquinone. Above 256°C appreciable decomposition occurred. Figure 17 compares both sets of experiments made with two data sets from the literature. Consistency below the start

of decomposition is reasonable. However, the instability of the compound upon heating causes greater scatter than usual. The sign and line rules were passed while the minimum in the $\Delta H/\Delta Z$ curve occurred at a reduced temperature of 0.84.

18. ε-Caprolactam - Previous experimental data were available for this compound only at very low temperatures. The data taken in this study are plotted together with the literature data in Figure 18 and show consistency over the entire range. unstable operation above 361°C prevented higher temperature results. The sign and line rules were passed, and a minimum in the $\Delta H/\Delta Z$ curve existed at a reduced temperature of 0.84.

19. N-Cyclohexylpyrrolidone - No literature data exist for this compound. Four sets of experimental data are plotted in Figure 19. Instabilities in taking data were more apparent with this compound. Decomposition began to occur near 350°C. As no reliable method exists for predicting the critical point, no qualification tests were run on the compound. However, the data when smoothed should be reliable and unconstrained regression coefficients are given.

20. Dimethyl sulfoxide - Four literature sources of data are compared with the two sets of data taken in this study in Figure 20. Excellent consistency exists over the entire temperature range. The sign and line rules were passed. At a reduced temperature of 0.84 a minimum in the $\Delta H/\Delta Z$ curve exists.

21. Trimethoxysilane - No literature data exist for this compound. A scarcity of sample dictated that data only be taken above atmospheric pressure. The two data sets are plotted in Figure 21 and are not so consistent as usual as the data using a higher boiling side-arm organic is lower than would be expected. Decomposition begins at the highest temperature. Regression of the data passes the sign rule with predicted criticals. The line rule was passed and a minimum in the $\Delta H/\Delta Z$ plot at a reduced temperature of 0.84 existed.

22. 3,3,3-Trifluoropropene - One set of proprietary data and the data taken in this work are plotted in Figure 22. Consistency is excellent (note that low temperature runs are at temperatures below room temperature) and data are smooth to the critical point. Regression was successful with data passing the sign and line rules.

ACKNOWLEDGMENTS

The authors acknowledge support of the Design Institute for Physical Property Data of the American Institute of Chemical Engineers for financial support of this work over the 1982-1984 period. Several other graduate and undergraduate students also contributed substantially to the completion of the work.

REFERENCES

Ambrose, D., "Equations for the Correlation and Estimation of Vapour Pressures," National Physical Laboratory Report Chem 114, United Kingdom, August 1980.

Chase, J. D., "The Qualification of Pure Component Physical Property Data," Chem. Eng. Progress 80(4) 63 (1984).

Daubert, T. E., Danner, R. P., "Data Compilation: Tables of Properties of Pure Compounds," AIChE, New York, extant 1985.

Hissong, D. W., M. S. Thesis, The Ohio State University, Columbus, Ohio (1965).

Hissong, D. W., Ph.D. Thesis, The Ohio State University, Columbus, Ohio(1968).

Kay, W. B., Personal Communication, April 20, 1983.

Klejnot, Olgierd J., Inorg. Chem., 2(4), 825-28 (1963).

Leonard, J. M. and J. D. Bultman, J. Chem. Ed., 33, 623, (1956).

McMicking, J. H., Ph.D. Thesis, The Ohio State University, Columbus, Ohio (1961).

Pak, S. C., M. S. Thesis, The Ohio State University, Columbus, Ohio (1969).

Smith, A. and A. W. C. Menzies, J. Am. Chem. Soc., 32, 897, 1412 (1910).

Spencer, C. F., M. S. Thesis, The Pennsylvania State University, University Park, Pennsylvania (1971).

Swietoslawski, W., "Ebulliometric Measurements," Reinhold, New York, 1945.

Swietoslawski, W., "Ebulliometry: The Physical Chemistry of Distillation," Chemical Publishing, New York, 1937.

Table 1
Experimental Data

	Compound	Experimental Vapor Pressure Data				
		T_c °C	P_c MPa	T_{range} °C	P_{range} Pa (10^{-5})	Notes
1.	Isopropyl Alcohol	234.21	4.716	65-234	0.49-47.2	
2.	Phenol	-	-	181-393	1.01-39.7	m
3.	n-Dodecylbenzene	-	-	326-391	0.96- 3.7	m
4.	p-Ethyltoluene	367.08	3.233	56-367	0.02-32.2	
5.	Methyl-t-butyl ether	223.25	3.397	24-223	0.32-34.0	
6.	Monoethanolamine	-	-	78-350	0.02-40.2	d
7.	Diethanolamine	-	-	188-244	0.07- 0.51	d
8.	2,4-Toluenediamine	-	-	193-355	0.17- 4.1	
9.	Hexamethylenediamine	-	-	122-291	0.09- 7.2	d
10.	Hexamethyleneimine	-	-	32-347	0.02-42.7	d
11.	Diethylene glycol	-	-	171-307	0.09- 9.7	
12.	Triethylene glycol	-	-	184-346	0.04- 3.1	
13.	Propylene glycol monomethyl ether acetate	-	-	56-256	0.03-11.3	
14.	Vinyl acetate	245.98	4.185	26-246	0.15-41.9	
15.	Dimethyl succinate	-	-	93-371	0.02-26.2	
16.	Ethyl acrylate	-	-	26-245	0.05-20.5	p
17.	Methacrylic acid	-	-	154-256	0.8 - 8.8	d
18.	ε-Caprolactam	-	-	171-361	0.1 - 9.7	
19.	N-Cyclohexylpyrrolidone	-	-	154-348	0.03- 2.7	d
20.	Dimethyl sulfoxide	-	-	118-254	0.11- 4.3	
21.	Trimethoxysilane	-	-	83-257	1.1 -44.0	s
22.	3,3,3-Trifluoropropene	105.44	3.609	-17-105	0.14- 3.6	

Notes: d = decomposition at highest temperature
p = polymerization at highest temperature
m = maximum temperature attainable in apparatus
s = insufficient sample for lower pressures

Table 2
Experimental Data in S.I. Units

1. Compound: Isopropyl Alcohol

T (C)	P (Pa)	Ref
65.26	48,680	9999
66.29	50,470	9999
66.75	52,190	9999
71.57	67,020	9999
75.28	74,740	9999
79.83	92,600	9999
85.27	117,420	9999
90.75	149,000	9999
96.67	181,680	9999
101.66	219,040	9999
105.21	244,560	9999
109.55	287,250	9999
113.70	328,600	9999
118.85	379,000	9999
123.52	431,580	9999
127.17	493,660	9999
130.56	529,500	9999
133.68	582,330	9999
139.90	688,380	9999
143.17	759,490	9999
145.96	798,900	9999
149.33	880,110	9999
151.66	944,300	9999
154.02	993,810	9999
95.72	179,330	9999
128.78	494,850	9999
135.51	592,320	9999
145.96	773,750	9999
152.38	932,940	9999
159.38	1,101,740	9999
174.68	1,584,690	9999
183.47	1,894,000	9999
186.83	1,954,410	9999
191.10	2,075,680	9999
198.33	2,438,760	9999
202.36	2,692,010	9999
205.55	2,897,040	9999
209.48	3,103,060	9999
216.09	3,494,500	9999
174.68	1,584,700	9999
183.47	1,894,000	9999
186.83	1,954,410	9999
191.10	2,075,680	9999
195.01	2,363,130	9999
198.33	2,438,760	9999
202.36	2,692,010	9999
205.55	2,897,040	9999
209.48	3,103,060	9999
216.09	3,494,500	9999
92.35	153,340	9999
109.55	271,850	9999
119.47	374,260	9999
134.18	579,140	9999
143.17	732,710	9999
157.88	1,055,050	9999
169.72	1,394,300	9999
176.49	1,618,320	9999
189.81	2,126,950	9999
199.89	2,580,260	9999
202.36	2,697,940	9999
209.48	3,097,480	9999
215.13	3,389,860	9999
219.70	3,695,770	9999
222.39	3,800,600	9999
223.37	3,954,240	9999
225.11	4,089,360	9999
217.29	3,519,400	9999
225.11	4,098,370	9999
231.39	4,530,760	9999
234.21	4,715,660	9999

2. Compound: Phenol

T (C)	P (Pa)	Ref
181.48	101,080	9999
191.49	129,750	9999
198.95	157,220	9999
209.08	202,200	9999
218.66	251,730	9999
227.08	301,410	9999
233.49	345,130	9999
242.74	419,640	9999
249.75	477,320	9999
254.72	529,490	9999
235.11	362,630	9999
243.17	421,280	9999
249.64	478,590	9999
255.53	547,810	9999
262.64	612,190	9999
262.53	610,840	9999
265.06	638,110	9999
270.09	698,590	9999
276.24	782,100	9999
282.19	859,630	9999
297.92	945,950	9999
293.07	1,028,600	9999
299.10	1,141,100	9999
302.46	1,201,600	9999
299.10	1,135,900	9999
305.19	1,238,100	9999
311.99	1,385,700	9999
322.90	1,637,800	9999
329.84	1,795,000	9999
336.64	1,963,700	9999
340.99	2,085,200	9999
187.74	116,980	9999
193.01	133,940	9999
198.21	148,780	9999
211.72	212,460	9999
218.03	247,630	9999
223.45	275,910	9999
229.84	318,060	9999
235.89	365,590	9999
315.33	1,432,600	9999
323.22	1,605,560	9999

T (C)	P (Pa)	Ref
330.53	1,774,450	9999
338.06	1,972,870	9999
344.82	2,162,990	9999
351.89	2,371,420	9999
357.21	2,545,910	9999
361.93	2,697,940	9999
367.01	2,925,570	9999
371.68	3,072,500	9999
377.30	3,315,110	9999
381.43	3,460,760	9999
385.96	3,667,420	9999
392.67	3,968,900	9999

3. Compound: n-Dodecyl benzene

T (C)	P (Pa)	Ref
342.85	137,360	9981
350.03	160,760	9981
355.83	177,770	9981
361.63	200,610	9981
367.22	227,890	9981
371.81	247,220	9981
376.35	270,200	9981
379.70	288,480	9981
383.59	308,540	9981
386.19	329,150	9981
390.72	366,160	9981
325.87	96,310	9980
329.66	105,080	9980
332.95	113,790	9980
335.64	120,570	9980
338.85	128,120	9980
342.01	136,290	9980
345.01	144,730	9980
347.70	154,670	9980
349.67	161,330	9980

4. Compound: para-Ethyl Toluene

T (C)	P (Pa)	Ref
163.04	104,120	9980
169.63	122,390	9980
179.58	155,900	9980
189.54	197,490	9980
199.52	245,550	9980
209.51	299,360	9980
209.64	300,630	9980
219.48	365,460	9980
229.47	435,800	9980
229.47	439,380	9980
239.46	520,390	9980
249.85	622,190	9980
256.32	620,360	9980
259.96	696,030	9980
270.04	735,520	9980
280.03	1,005,440	9980
290.02	1,172,160	9980
300.03	1,351,600	9980
310.02	1,361,850	9980
320.34	1,571,990	9980

T (C)	P (Pa)	Ref
330.00	2,053,950	9980
340.00	2,343,050	9980
350.00	2,653,520	9980
350.37	2,645,860	9980
360.00	3,006,060	9980
367.08	3,233,410	9980
151.64	76,680	9950
142.55	59,970	9950
130.25	41,740	9950
123.01	33,320	9950
116.76	27,400	9950
102.73	16,980	9950
93.37	12,090	9950
83.78	8,390	9950
74.81	5,930	9950
67.48	4,130	9950
88.44	9,940	9951
83.37	8,090	9951
77.64	6,370	9951
70.24	4,660	9951
61.84	3,210	9951
56.12	2,370	9951

5. Compound: Methyl-t-Butyl Ether

T (C)	P (Pa)	Ref
55.85	104,000	9980
78.35	213,000	9980
86.75	259,000	9980
88.75	272,000	9980
91.75	296,000	9980
91.75	286,000	9980
95.25	321,000	9980
98.25	351,000	9980
102.05	380,000	9980
103.75	394,000	9980
105.05	401,000	9980
107.55	430,000	9980
108.85	443,000	9980
110.05	454,000	9980
111.75	478,000	9980
114.65	504,000	9980
115.95	523,000	9980
117.15	532,000	9980
119.45	563,000	9980
119.95	564,000	9980
121.25	585,000	9980
123.25	617,000	9980
123.75	610,000	9980
125.55	648,000	9980
127.35	664,000	9980
127.85	688,000	9980
130.25	706,000	9980
130.65	727,000	9980
133.15	749,000	9980
135.15	798,000	9980
137.65	817,000	9980
140.25	866,000	9980
143.65	924,000	9980
146.45	988,000	9980

T (C)	P (Pa)	Ref
149.25	1,024,000	9980
152.05	1,043,000	9980
154.75	1,120,000	9980
155.95	1,162,000	9980
170.45	1,378,000	9980
174.05	1,480,000	9980
181.95	1,678,000	9980
186.55	1,906,000	9980
189.65	1,961,000	9980
194.85	2,167,000	9980
198.65	2,291,000	9980
204.45	2,436,000	9980
205.75	2,551,000	9980
210.05	2,704,000	9980
213.15	2,883,000	9980
223.25	3,397,000	9980
44.46	70,340	9950
42.20	65,000	9950
37.02	53,380	9950
34.53	48,370	9950
32.19	44,010	9950
29.13	39,660	9950
26.20	34,430	9950
24.47	32,210	9950

6. Compound: Monoethanolamine

T (C)	P (Pa)	Ref
191.19	203,020	9995
155.24	78,230	9995
167.73	104,660	9995
172.92	123,480	9995
177.38	138,780	9995
183.03	160,330	9995
170.01	134,610	9999
180.02	171,290	9999
190.01	222,780	9999
200.00	281,240	9999
210.01	344,260	9999
220.02	429,690	9999
230.02	539,500	9999
240.01	642,430	9999
250.00	775,490	9999
260.01	925,140	9999
270.00	1,107,590	9999
280.01	1,349,980	9999
290.01	1,638,380	9999
300.01	1,848,240	9999
320.01	2,518,550	9999
330.02	2,898,080	9999
340.01	3,395,280	9999
350.02	4,023,590	9999
160.36	76,890	9950
156.48	65,580	9950
148.83	50,690	9950
139.08	35,650	9950
129.17	24,830	9950
122.94	17,630	9950
117.76	14,790	9950

T (C)	P (Pa)	Ref
104.44	8,010	9950
95.20	5,250	9950
86.26	3,150	9950
78.34	2,030	9950

7. Compound: Diethanolamine

T (C)	P (Pa)	Ref
188.13	7,040	9950
208.43	15,280	9950
221.76	24,340	9950
230.76	32,970	9950
237.13	41,000	9950
244.00	50,800	9950

8. Compound: 2,4-Toluenediamine

T (C)	P (Pa)	Ref
301.229	141,680	9980
318.323	236,840	9980
326.839	274,620	9980
334.859	335,120	9980
345.329	347,830	9980
353.246	410,710	9980
219.07	17,400	9950
225.22	20,500	9950
232.20	25,130	9950
192.66	6,610	9951
208.11	11,350	9951
221.07	17,130	9951
227.60	20,760	9951
243.62	34,880	9951
252.37	42,360	9951

9. Compound: Hexamethylenediamine

T (C)	P (Pa)	Ref
200.85	100,000	9980
204.25	111,000	9980
207.65	121,000	9980
212.35	137,000	9980
214.35	145,000	9980
217.95	157,000	9980
222.35	174,000	9980
223.85	185,000	9980
225.75	191,000	9980
230.05	213,000	9980
230.05	207,000	9980
235.45	242,000	9980
240.15	267,000	9980
244.15	294,000	9980
248.45	319,000	9980
252.45	347,000	9980
255.65	374,000	9980
258.45	395,000	9980
261.56	425,000	9980
264.35	446,000	9980

T (C)	P (Pa)	Ref
267.95	473,000	9980
275.75	545,000	9980
279.65	584,000	9980
283.75	622,000	9980
287.25	662,000	9980
290.65	716,000	9980
177.42	56,630	9950
173.12	49,810	9950
166.16	40,690	9950
183.91	67,650	9950
185.04	69,660	9950
145.26	20,910	9950
142.48	18,880	9950
141.61	18,330	9950
140.06	17,180	9950
138.72	16,480	9950
133.90	13,670	9950
132.27	12,990	9950
127.20	10,710	9950
121.69	8,650	9950

10. Compound: Hexamethyleneimine

T (C)	P (Pa)	Ref
157.95	168,000	9980
160.05	177,000	9980
162.25	191,000	9980
164.15	203,000	9980
167.35	213,000	9980
169.95	230,000	9980
173.45	250,000	9980
176.15	264,000	9980
186.85	337,000	9980
190.35	347,000	9980
194.45	380,000	9980
199.15	410,000	9980
203.15	443,000	9980
206.15	475,000	9980
209.45	506,000	9980
212.25	533,000	9980
215.05	557,000	9980
217.95	578,000	9980
220.55	608,000	9980
223.55	646,000	9980
227.25	691,000	9980
231.25	724,000	9980
235.15	790,000	9980
238.45	830,000	9980
241.65	874,000	9980
244.95	942,000	9980
247.95	994,000	9980
250.65	1,034,000	9980
253.15	1,082,000	9980
255.95	1,134,000	9980
271.15	1,337,000	9980
275.05	1,513,000	9980
279.95	1,629,000	9980
285.55	1,783,000	9980
289.05	1,911,000	9980
292.85	2,033,000	9980
296.95	2,162,000	9980
300.75	2,277,000	9980
303.85	2,416,000	9980
306.55	2,493,000	9980
311.95	2,684,000	9980
317.05	2,848,000	9980
322.55	3,075,000	9980
326.25	3,247,000	9980
330.25	3,406,000	9980
333.65	3,601,000	9980
337.35	3,762,000	9980
340.35	3,938,000	9980
343.55	4,110,000	9980
346.65	4,272,000	9980
127.53	74,300	9950
122.78	64,490	9950
118.92	57,980	9950
102.97	34,260	9950
96.42	27,510	9950
90.71	22,270	9950
84.59	17,650	9950
76.50	12,700	9950
71.62	10,260	9950
67.61	8,890	9950
64.35	7,460	9950
58.87	5,720	9950
53.97	4,400	9950
53.01	4,180	9950
50.83	3,680	9950
45.97	3,310	9950
39.09	2,330	9950
32.29	1,590	9950

11. Compound: Diethylene Glycol

T (C)	P (Pa)	Ref
171.19	9,510	9952
184.33	15,650	9952
223.40	75,230	9952
192.30	29,890	9952
200.26	27,500	9952
208.87	36,350	9952
216.20	45,550	9952
221.28	42,940	9952
226.16	61,750	9952
230.13	68,530	9952
171.35	9,480	9952
178.96	12,890	9952
188.30	18,110	9952
227.69	64,390	9952
231.56	71,450	9952
197.17	24,430	9952
205.70	32,620	9952
212.89	41,090	9952
219.03	49,920	9952
223.69	57,000	9952
306.83	368,060 *	9980
318.36	460,800 *	9980

T (C)	P (Pa)	Ref
328.14	548,560 *	9980
335.72	613,700 *	9980
343.07	698,700 *	9980
354.87	910,350 *	9980
361.21	968,730 *	9980
235.94	80,590	9981
239.62	88,940	9981
243.83	94,610	9981
248.63	114,300	9981
254.72	132,910	9981
260.10	154,540	9981
267.45	200,740	9981
274.02	234,220	9981
278.51	271,030	9981
282.44	292,040	9981
288.52	322,550	9981
293.97	354,940	9981
296.26	385,820	9981

12. Compound: Triethylene Glycol

T (C)	P (Pa)	Ref
284.14	102,280 *	9999
288.43	110,260 *	9999
294.44	122,440 *	9999
295.98	123,290 *	9999
300.90	134,140 *	9999
304.90	147,800 *	9999
310.49	161,820 *	9999
315.78	179,350 *	9999
320.64	204,880 *	9999
327.32	233,190 *	9999
331.83	252,550 *	9999
336.56	271,520 *	9999
341.51	285,890 *	9999
345.97	311,250 *	9999
209.09	10,680	9950
210.22	10,830	9950
223.00	16,820	9950
183.68	4,230	9951
191.63	5,900	9951
211.91	11,720	9951
218.09	14,610	9951
224.04	17,440	9951
228.40	20,080	9951
232.24	22,550	9951
236.65	25,620	9951
241.18	29,620	9951
240.16	29,350	9952
243.54	32,790	9952
246.55	36,370	9952
250.45	40,310	9952
256.55	46,130	9952
262.11	54,080	9952
264.65	57,250	9952
266.58	60,670	9952

13. Compound: Propylene Glycol MEA

T (C)	P (Pa)	Ref
186.64	275,400	9980
191.88	316,000	9980
196.75	351,730	9980
201.31	393,450	9980
204.81	422,020	9980
208.54	458,730	9980
212.07	495,160	9980
215.70	531,630	9980
219.43	569,890	9980
223.71	622,510	9980
228.29	677,200	9980
232.23	738,820	9980
235.66	782,909	9980
239.45	832,450	9980
242.42	888,690	9980
254.80	943,400	9980
248.12	994,060	9980
251.00	1,037,770	9980
253.47	1,078,920	9980
256.16	1,131,680	9980
255.95	1,141,590	9980
262.38	1,283,990	9980
267.68	1,406,170	9980
272.73	1,513,760	9980
276.94	1,691,140	9980
282.71	1,785,390	9980
286.82	1,896,850	9980
292.08	2,073,120	9980
295.94	2,205,070	9980
302.42	2,495,790	9980
307.44	2,769,130	9980
312.84	3,104,590	9980
317.45	3,567,310	9980
160.63	122,860	9980
169.74	160,630	9980
176.17	221,580	9980
180.20	246,000	9980
185.38	274,930	9980
190.33	313,400	9980
194.93	348,360	9980
199.46	383,720	9980
203.15	416,990	9980
207.25	460,710	9980
211.30	495,010	9980
215.35	534,640	9980
56.35	3,070	9952
73.15	7,110	9952
83.35	11,220	9952
90.65	15,100	9952
95.95	18,950	9952
100.85	22,880	9952
105.25	26,940	9952
109.15	30,880	9952
112.55	34,800	9952
115.55	38,660	9952
117.85	42,040	9952

T (C)	P (Pa)	Ref
120.75	46,290	9952
123.45	50,880	9952
125.65	54,720	9952
127.75	58,900	9952
129.95	62,480	9952
131.95	66,730	9952
133.85	70,490	9952
135.55	74,410	9952
137.45	78,650	9952

14. Compound: Vinyl Acetate

T (C)	P (Pa)	Ref
160.28	992,970	9980
170.02	1,164,430	9980
181.95	1,426,970	9980
199.73	1,639,080	9980
206.80	1,969,600	9980
211.69	2,400,440	9980
216.81	2,482,050	9980
220.52	2,760,590	9980
224.96	2,968,110	9980
121.69	4,084,010	9980
119.09	392,170	9980
113.80	350,940	9980
108.57	312,650	9980
101.55	261,970	9980
93.48	217,750	9980
126.20	490,970	9980
130.26	524,900	9980
134.04	576,530	9980
140.39	628,540	9980
144.20	711,760	9980
186.17	156,520	9980
218.32	253,890	9980
227.84	301,720	9980
231.52	3,185,410	9980
239.10	3,465,860	9980
67.08	86,340	9980
86.10	147,080	9980
96.17	214,210	9980
104.12	260,600	9980
111.73	319,490	9980
117.26	368,210	9980
123.43	422,280	9980
127.89	455,880	9980
132.52	511,620	9980
135.98	571,830	9980
139.57	620,640	9980
143.57	670,820	9980
146.22	712,420	9980
148.77	754,900	9980
150.81	787,370	9980
153.14	800,870	9980
155.17	842,610	9980
200.84	2,049,390	9980
206.59	2,253,520	9980
214.23	2,389,490	9980
218.61	2,539,920	9980
223.46	2,821,520	9980

T (C)	P (Pa)	Ref
226.78	3,005,720	9980
231.88	3,304,780	9980
234.77	3,415,960	9980
237.50	3,551,630	9980
239.81	3,666,960	9980
241.98	3,800,610	9980
245.98	4,185,380	9980
218.61	2,539,920	9980
223.46	2,821,520	9980
226.78	3,005,720	9980
231.88	3,304,780	9980
234.77	3,415,960	9980
237.50	3,551,630	9980
239.81	3,666,960	9980
241.98	3,800,610	9980
245.98	4,185,380	9980
64.08	75,140	9950
57.08	56,980	9950
48.96	42,210	9950
43.38	32,990	9950
52.21	47,410	9950
42.84	32,020	9950
33.24	20,910	9950
25.59	14,740	9950

15. Compound: Dimethyl Succinate

T (C)	P (Pa)	Ref
93.59	2,430	9952
117.40	7,270	9952
131.41	12,750	9952
141.43	18,760	9952
147.16	23,170	9952
152.66	27,910	9952
158.25	33,490	9952
162.06	37,740	9952
166.39	43,830	9952
167.59	43,910	9952
171.87	51,220	9952
176.33	58,820	9952
180.36	65,530	9952
184.33	73,120	9952
185.02	73,430	9952
186.79	77,810	9952
265.58	453,300	9980
273.33	537,890	9980
278.57	596,630	9980
283.82	652,700	9980
288.64	714,530	9980
292.44	758,650	9980
295.91	810,270	9980
298.97	858,270	9980
302.75	904,110	9980
305.84	954,450	9980
308.10	1,001,630	9980
313.38	1,079,440	9980
317.79	1,164,415	9980
322.55	1,257,920	9980
326.42	1,344,070	9980
330.25	1,417,250	9980

T (C)	P (Pa)	Ref
334.70	1,441,500	9981
339.01	1,626,210	9981
344.40	1,745,780	9981
350.53	1,918,470	9981
356.63	2,100,760	9981
361.59	2,256,690	9981
366.22	2,399,040	9981
371.45	2,615,200	9981
202.77	95,770	9982
206.89	108,660	9982
210.19	118,310	9982
213.20	129,620	9982
217.21	146,490	9982
222.74	159,610	9982
227.52	177,440	9982
231.14	201,670	9982
235.13	224,130	9982
238.10	236,760	9982
240.95	261,790	9982
241.13	292,800	9982
247.42	313,650	9982
251.55	340,820	9982
254.58	371,110	9982

16. Compound: Ethyl Acrylate

T (C)	P (Pa)	Ref
100.75	110,000	9980
104.95	123,000	9980
108.75	137,000	9980
111.75	152,000	9980
115.05	166,000	9980
117.55	179,000	9980
119.65	188,000	9980
122.05	201,000	9980
125.85	223,000	9980
129.85	246,000	9980
133.35	264,000	9980
136.36	288,000	9980
139.55	306,000	9980
141.95	325,000	9980
144.85	351,000	9980
147.65	378,000	9980
149,65	399,000	9980
151.65	421,000	9980
168.05	588,000	9980
179.55	743,000	9980
190.45	911,000	9980
197.85	1,032,000	9980
203.85	1,151,000	9980
208.75	1,263,000	9980
213.15	1,374,000	9980
217.25	1,452,000	9980
228.65	1,768,000	9980
237.75	1,904,000	9980 p
244.85	2,047,000	9980 p
89.08	70,310	9950
83.69	58,070	9950
75.72	44,060	9950
71.40	37,250	9950
68.19	32,730	9950
61.54	25,000	9950
54.78	18,520	9950
47.54	13,760	9950
41.90	11,200	9950
35.26	8,240	9950
38.16	9,470	9950
31.88	6,920	9950
28.63	5,860	9950
26.53	5,230	9950

17. Compound: Methacrylic Acid

T (C)	P (Pa)	Ref
154.34	79,390	9981
164.60	111,540	9981
173.51	140,140	9981
184.22	181,510	9981
196.75	253,810	9981
203.32	298,440	9981
209.86	350,560	9981
215.69	397,220	9981
180.83	165,340	9982
192.76	242,960	9982
200.94	293,800	9982
207.96	350,110	9982
213.57	400,930	9982
219.44	472,040	9982
224.62	485,760	9982
230.97	557,790	9982
235.16	598,460	9982
240.22	662,880	9982
244.87	716,810	9982
248.10	757,170	9982
252.06	830,510	9982
256.38	882,140	9982

18. Compound: Epsilon Caprolactam

T (C)	P (Pa)	Ref
275.28	116,250	9980
279.12	123,540	9980
288.52	151,550	9980
294.77	172,270	9980
299.03	185,260	9980
304.35	207,660	9980
309.50	229,590	9980
314.33	243,030	9980
319.58	272,280	9980
329.62	326,740	9980
340.27	409,090	9980
354.01	501,110	9980
361.47	447,330	9980
371.18	678,590	9980
182.80	8,170	9952
194.71	12,920	9952
217.00	26,630	9952
223.60	32,560	9952
229.79	37,760	9952

T (C)	P (Pa)	Ref
236.61	45,880	9952
241.81	52,260	9952
246.09	58,590	9952
250.71	65,620	9952
254.89	72,780	9952
258.09	78,220	9952
208.80	20,310	9952
203.20	17,030	9952

19. Compound: n-Cyclohexylpyrrolidone

T (C)	P (Pa)	Ref
263.78	53,480	9980
272.80	69,370	9980
284.13	88,470	9980
292.44	93,710	9980
297.27	110,120	9980
300.55	116,020	9980
304.46	131,770	9980
308.58	140,240	9980
312.32	144,350	9980
317.08	160,540	9980
320.81	169,980	9980
326.83	118,570	9980
329.98	191,750	9980
334.29	218,350	9980
337.66	216,370	9980
341.02	246,390	9980
343.71	256,170	9980
347.70	268,890	9980
268.44	61,230	9980
275.86	74,170	9980
281.49	83,680	9980
295.84	105,020	9980
302.15	122,240	9980
309.82	142,870	9980
315.49	154,990	9980
321.09	170,800	9980
327.44	189,100	9980
331.23	199,130	9980
333.82	215,310	9980
342.49	241,710	9980
345.83	262,860	9980
274.68	73,370	9981
277.85	77,750	9981
282.21	85,510	9981
290.32	91,580	9981
292.11	102,970	9981
294.51	108,740	9981
296.65	114,540	9981
299.04	113,790	9981
301.77	118,860	9981
305.00	134,780	9981
307.93	138,670	9981
312.73	144,740	9981
153.53	3,660	9950
174.90	6,650	9950
191.50	10,550	9950
203.75	14,240	9950
212.24	17,160	9950

T (C)	P (Pa)	Ref
218.22	20,110	9951
225.94	25,730	9951
233.15	31,250	9951
233.31	31,470	9951
238.84	36,100	9951
244.35	41,080	9951
249.03	46,070	9951
254.07	51,190	9951
257.84	55,560	9951

20. Compound: Dimethyl Sulfoxide

T (C)	P (Pa)	Ref
118.95	10,960	9952
127.45	14,980	9952
134.15	18,850	9952
139.65	22,730	9952
144.65	26,920	9952
148.95	30,780	9952
152.75	34,640	9952
156.65	38,860	9952
160.25	46,720	9952
165.34	50,540	9952
168.15	54,720	9952
170.55	58,620	9952
172.95	62,530	9952
175.25	66,670	9952
177.35	70,450	9952
191.08	100,370	9980
194.83	109,710	9980
198.07	119,490	9980
201.98	132,890	9980
206.09	146,100	9980
209.76	159,050	9980
214.42	175,010	9980
217.93	191,910	9980
223.47	216,220	9980
227.53	238,150	9980
232.21	261,250	9980
236.58	289,070	9980
241.25	318,370	9980
245.57	348,210	9980
249.19	380,620	9980
254.22	429,410	9980

21. Compound: Trimethoxysilane

T (C)	P (Pa)	Ref
83.54	110,000	9980
89.23	124,550	9980
94.58	146,660	9980
98.66	164,640	9980
102.46	187,020	9980
106.27	207,320	9980
109.23	224,180	9980
112.11	243,690	9980
115.48	264,850	9980
118.33	278,940	9980
121.66	309,910	9980
125.39	342,810	9980

T (C)	P (Pa)	Ref
129.86	375,680	9980
133.34	410,330	9980
139.22	461,640	9980
142.38	503,030	9980
144.77	526,770	9980
147.65	572,030	9980
150.02	598,510	9980
152.85	634,770	9980
154.39	661,950	9980
166.54	708,770	9981
176.78	893,000	9981
185.05	990,790	9981
189.62	1,077,780	9981
194.83	1,245,390	9981
199.35	1,372,490	9981
203.55	1,468,410	9981
208.32	1,598,650	9981
212.44	1,729,100	9981
216.44	1,838,610	9981
220.17	1,920,740	9981
224.97	2,078,650	9981
229.35	2,250,470	9981
233.30	2,449,660	9981
236.94	2,663,340	9981
239.89	2,752,070	9981
242.79	2,985,450	9981
246.00	3,126,080	9981
248.66	3,280,510	9981
251.35	3,466,070	9981
257.24	4,395,670	9981

22. Compound: 3,3,3-Trifluoropropene

T (C)	P (Pa)	Ref
105.44	3,608,900	9975
100.57	3,312,600	9975
98.03	3,185,200	9975
94.40	3,006,300	9975
93.97	2,974,500	9975
91.68	2,781,700	9975
89.60	2,664,500	9975
88.37	2,655,600	9975
85.69	2,538,800	9975
81.32	2,254,800	9975
80.00	2,182,700	9975
81.47	2,251,800	9975
78.94	2,199,400	9975
76.90	2,107,000	9975
74.52	1,964,400	9975
72.82	1,941,800	9975
70.28	1,783,900	9975
67.58	1,687,400	9975
64.75	1,592,500	9975
62.15	1,507,500	9975
59.06	1,395,100	9975
56.82	1,316,400	9975
55.06	1,269,600	9975
54.21	1,219,100	9975
53.93	1,237,300	9975
52.82	1,185,500	9975
50.69	1,143,600	9975
48.94	1,074,100	9975
48.38	1,069,300	9975
44.11	933,200	9975
41.89	903,000	9975
39.04	825,900	9975
36.78	765,400	9975
29.25	641,600	9975
0.38	274,000	9975
-3.61	236,200	9975
-12.79	173,600	9975
-15.41	146,300	9975
-17.26	141,200	9975

* data suspect
p polymerizing

Table 3
Regression Coefficients and Critical Properties

Compound	Regression Coefficients A	B	C	D	E	TC °C	PC MPa	Notes		Units
1. Isopropyl Alcohol	106.403	- 8460.465	-13.547	$1.2(10^{-5})$	2.0	234.2	4.716	b		kPa
2. Phenol	133.9634	-10754.354	-18.4321	0.015615	1.0	421E	6.13E	a		kPa
3. n-Dodecylbenzene	171.0771	-14777.7085	-22.71175	0.0172567	1.0	501E	1.58P	a		Pa
4. p-Ethyltoluene	122.3845	- 9169.6999	-17.0396	0.01567	1.0	367.1	3.233	b		kPa
5. Methyl-t-butyl ether	60.4976	- 5632.35	- 6.675	$2.81(10^{-17})$	6.0	223.3	3.397	b		kPa
6. Monoethanolamine	113.8054	-11035.8680	-14.0779	$8.(10^{-6})$	2.0	365P	6.87P	a	L	kPa
7. Diethanolamine	195.1156	-15859.054	-26.393	0.02171	1.0	442P	3.27P	a	D	Pa
8. 2,4-Toluenediamine	112.641	-12510.339	-12.663	$4.517(10^{-6})$	2.0	531P	4.38P	a	L	Pa
9. Hexamethylenediamine	86.988	- 9967.57	- 8.847	$6.50(10^{-18})$	6.0	390P	3.29P	a	L	Pa
10. Hexamethyleneimine	66.921	- 7575.46	- 7.299	$1.01(10^{-17})$	6.0	342P	4.27P	a	L	kPa
11. Diethylene glycol	153.034	-15227.3358	-18.29816	$8.7(10^{-6})$	2.0	407P	4.60P	a	LC	Pa
12. Triethylene glycol	161.9005	-17519.38	-20.3711	$9.6(10^{-6})$	2.0	427P	3.32P	a	LC	kPa
13. Propylene glycol monomethyl ether acetate	156.1846	-10467.51	-21.3296	0.0214855	1.0	311P	3.07P		L	Pa
14. Vinyl acetate	64.0782	- 6394.1288	- 7.0141	$1.92(10^{-17})$	6.0	246.0	4.185	b		kPa
15. Dimethyl succinate	160.103	-11780.901	-21.566	0.019515	1.0	377P	3.21P		L	Pa
16. Ethyl acrylate	122.445	- 8084.71	-17.394	0.0185	1.0	280P	3.68P	a	L	kPa
17. Methacrylic acid	98.0824	- 9486.580	-12.0074	$6.7(10^{-6})$	2.0	370P	4.70P	a	L	kPa
18. ε-Caprolactam	106.216	-11703.178	-11.8127	$4.20(10^{-6})$	2.0	533P	4.77P		L	Pa
19. N-Cyclohexylpyrrolidone	5272.6	-245000.	-825.31	0.71226	1.0	----	----	U		Pa
20. Dimethyl sulfoxide	130.9049	-10289.754	-16.866	0.013668	1.0	453P	5.65P	a	L	Pa
21. Trimethoxysilane	98.98085	- 7285.1785	-11.6256	$1.03(10^{-5})$	2.0	337P	2.97P		L	Pa
22. 3,3,3-Trifluoropropene	84.912	- 4586.748	-10.163	$1.85(10^{-5})$	2.0	105.4	3.609	b		Pa

a = T_c, P_c from DIPPR Compilation
C = use coefficients with caution
E = reported experimental
P = Predicted

b = T_c, P_c from this work
D = predicted by Dow
L = Lydersen Method for T_c and P_c
U = unconstrained fit

Pa = pascals
kPa = kilopascals

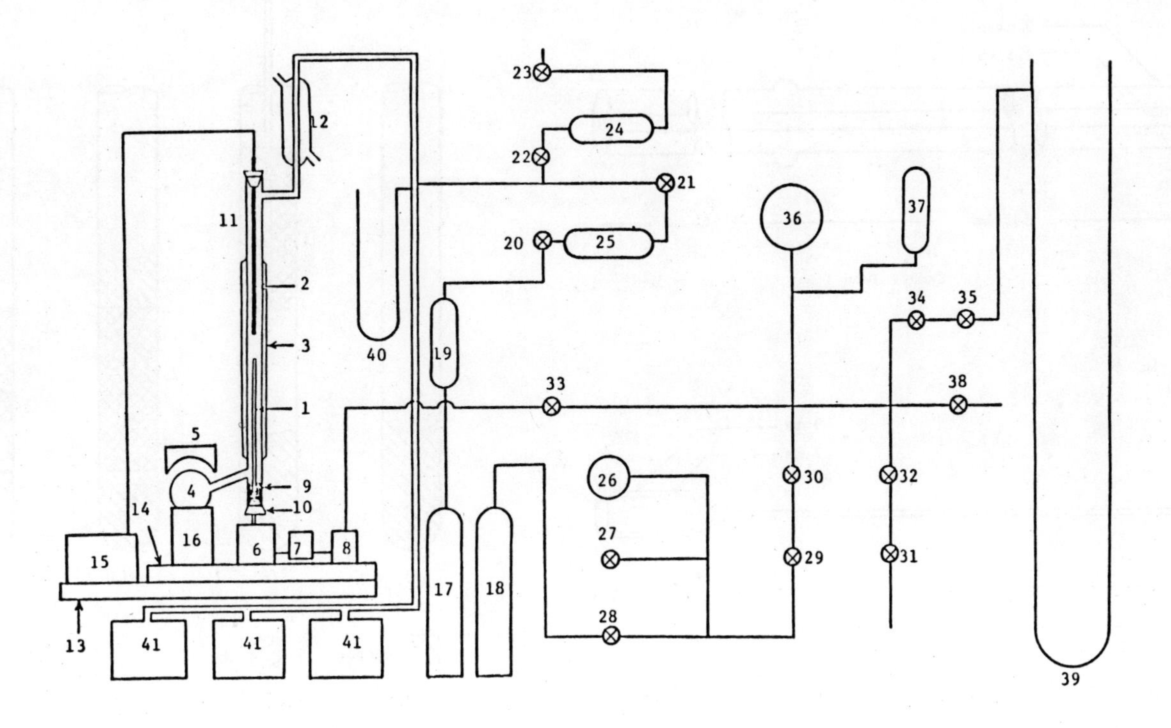

1. Experimental tube
2. Copper constantan thermocouple
3. Double walled vacuum heating jacket
4. Side arm boiler
5. Upper heating mantle
6. Front Hg leg
7. Compressor block assembly
8. Rear Hg leg
9. Hg layer
10. Rubber stopper
11. Cooling air stream
12. Water condenser
13. Supporting platform
14. Hg spill tray
15. Potentiometer
16. Electric heater
17. Inert gas tank
18. Nitrogen tank
19. Surge tank

20-23. Needle valves (pressure regulating device)

24. Air line pressure adjusting bulb
25. Vacuum line pressure adjusting bulb
26. Pressure gage
27. Release valve
28. Gas line valve

29-30. System inlet valves

31-32. System exhaust valves

33. Compressor inlet valve
34. Pressure check valve
35. Release valve
36. Heise gage
37. Surge tank
38. Valve for Kuska apparatus
39. Open ended manometer
40. Closed ended manometer
41. Five-gallon containers

Figure A. High pressure vapor pressure apparatus.

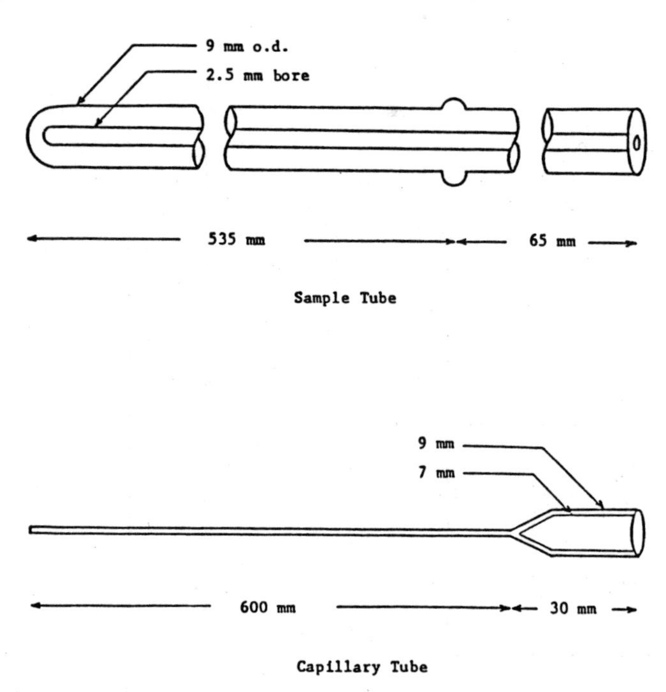

Figure B. Diagram of sample and capillary tubes.

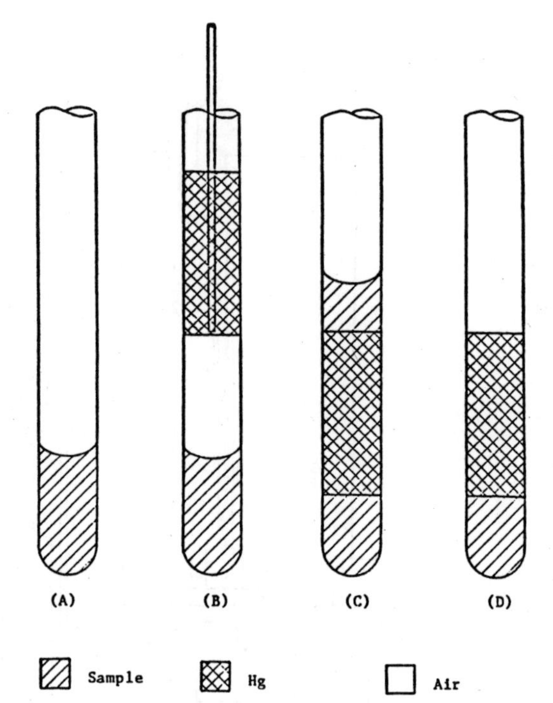

Figure D. Sample tube loading sequence.

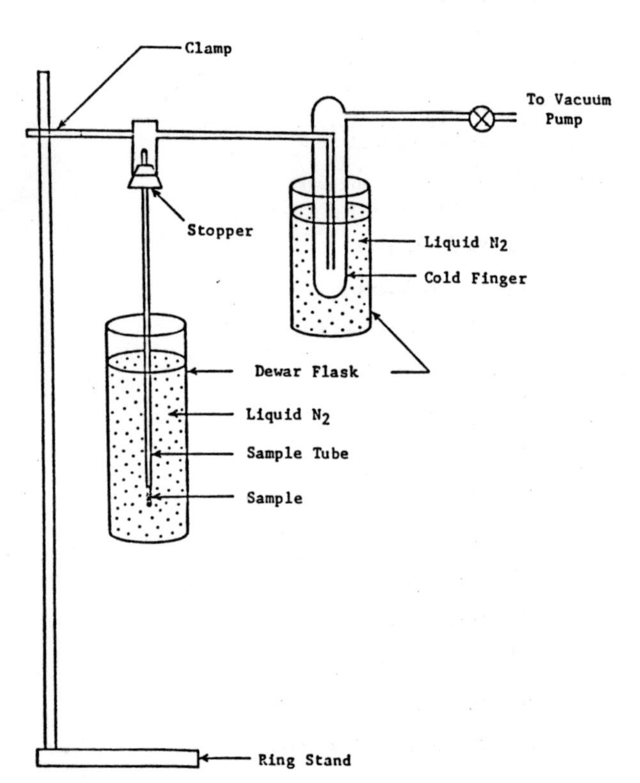

Figure C. Degassing apparatus.

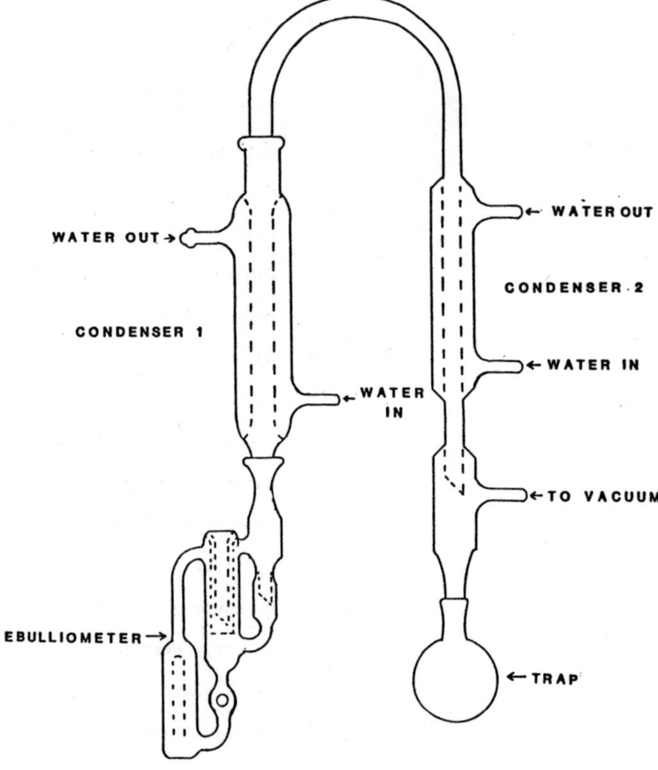

Figure E. Low pressure vapor pressure apparatus.

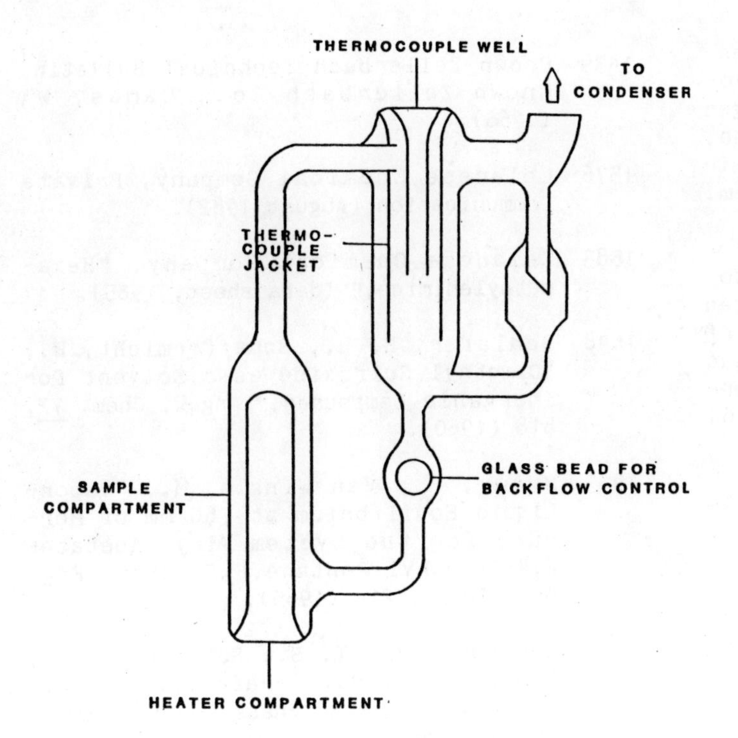

Figure F. Ebulliometer.

References for Figures 1-22

8 Kirk-Othmer, "Encyclopedia of Chemical Technology," 2nd ed., (22 Vol. + Suppl.), Interscience, New York (1966).

20 Stull, D. R., "Vapor Pressure of Pure Substances," Ind. Eng. Chem. 39, 517 (1947).

30 Wilhoit, R. C., Zwolinski, B. J., "Physical and Thermodynamic Properties of Aliphatic Alcohols," J. Phys. Chem. Ref. Data 2 (Suppl. No. 1) (1973).

39 Riddick, J. A., Bunger, W. B., "Organic Solvents: Physical Properties and Methods of Purification," 3rd ed., Wiley Interscience, New York (1970).

41 Ambrose, D., Sprake, C. H. S., "Vapor Pressure of Alcohols," J. Chem. Thermo. 2, 631 (1970).

47 Thermodynamic Research Center, "Selected Values of Properties of Hydrocarbons and Related Compounds," American Petroleum Institute Research Project 44, Texas A&M University, College Station, Texas (loose-leaf data sheets, extant) (1980).

50 Timmermans, J., "Physico-Chemical Constants of Pure Organic Substances (2 vols.)," 2nd ed., Elsevier, New York (1965).

57 Forziati, A. F., Norris, W. R., Rossini, F. D., "Vapor Pressures and Boiling Points of Sixty API-NBS Hydrocarbons," J. Res. Natl. Bur. Std. A 43, 555 (1949).

177 Boublik, T., Fried, V., Hala, E., "The Vapour Pressures of Pure Substances," Elsevier, New York (1973).

406 Kudchadker, A. P., Kudchadker, S. A., Wilhoit, R. C., "Key Chemicals Data Books - Phenol," Thermodynamics Research Center, Texas Engineering Experiment Station, Texas A&M University, College Station Texas (1977).

409 Rinkenbach, W. H., "Properties of Diethylene Glycol," Ind. Eng. Chem. 19, 1974 (1927).

422 Kirk-Othmer, "Encyclopedia of Chemical Technology," 3rd ed., Interscience, New York (1978).

460 Marsden, J., Cuthbertson, A. C., "Vapor Pressure of Vinyl Acetate," Can. J. Res. 9, 419 (1933).

467 Morrison, G. O., Shaw, T. P. G., "By-Products of the Carbide Industry: The Manufacture of Ethylidine Diacetate and Vinyl Acetate," Trans. Electrochem. Soc. 63, 425 (1933).

511 Tennessee Eastman Vapor Pressure Data.

531 McDonald, R. A., Shrader, S. H., Stull, D. P., "Vapor Pressure and Freezing Points of Thirty Organics," J. Chem. Eng. Data 4(4), 311 (1959).

533 Ambrose, D., Townsend R., "Thermodynamic Properties of Organic Oxygen Compounds. Part IX. The Critical Properties and Vapour Pressures Above Five Atmospheres of Six Aliphatic Alcohols," J. Chem. Soc. 3614 (1963).

543 Stage, H., et al., "Erdol U. Kohle, 6(7), 375 (1953).

934 Ambrose, D., Counsell, J. F., Laurenson, I. J., Lewis, G. B., "Thermodynamic Properties of Organic Oxygen Compounds. XLVII. Pressure, Volume, Temprature Relations and Thermodynamic Properties of Propan-2-ol," J. Chem. Thermo 10, 1033 (1978).

937 Ambrose, D., Hall, D. J., "Thermodynamic Properties of Organic Oxygen Compounds. L. The Vapour Pressures of 1,2-Ethanediol (Ethylene Glycol) and bis(2-Hydroxyethyl) ether (Diethylene Glycol)," J. Chem. Thermo. 13, 61 (1981).

940 Riddle, E. H., "Monomeric Acrylic Esters," Reinhold Publishing Corp., New York (1954).

941 Celanese Product Bulletin, "Ethyl Acrylate," Celanese Chemical Company, Inc., Dallas, Texas.

1101 Petro-Tex Chemical Corp., "Methyl tert-Butyl Ether," (material safety data sheet), Houston, Texas (1979).

1199 Hawley, G G., "The Condensed Chemical Dictionary," 9th edition, Van Nostrand Reinhold, Co., New York (1977).

1257 Riedel, L., "Eine Neue Universelle Damfdruck-formal," Chem. Ing. Tech. 26, 83 (1954).

1262 Maxwell, J. B., Bonnell, L. S., "Vapor Pressure Charts for Petroleum Engineers," Esso Research and Engineering Company, Linden, New Jersey (1955).

1293 Verschueren, K., "Handbook of Environmental Data on Organic Chemicals," Van Nostrand Reinhold, New York (1977).

1439 Crown Zellerbach Technical Bulletin, Crown Zellerbach Co., Camas, WA (1966).

1576 Celanese Chemical Company, Private Communication (August 1982).

1583 Celanese Chemical Company, "Hexamethylenimine," (data sheet, 1965).

1648 Schlafer, H. L., Schaffermicht, W., "Dimethyl Sulfoxide as a Solvent for Inorganic Compounds," Angew. Chem. 72, 618 (1960).

1973 Swamy, P., Van Winkle, M., "Vapor-Liquid Equilibrium at 760 mm of Mercury for the System Vinyl Acetate-2,4-Dimethyl Pentane," J. Chem. Eng. Data 10(3), 214 (1965).

1974 Rudakovskaya, T. S., Soboleva, S. A., Temofeev, V. S., Serafermov, L. A., "Investigation of Phase Equilibrium in the Ternary System Vinyl Acetate-Methanol-Cyclohexane," J. Appl. Chem. USSR 41, 1479 (1968).

1981 Ambrose, D., Ellender, J. H., Sprake, H. S., Townsend, R., XLIII. "Thermodynamic Properties of Organic Compounds - XLIII. Vapour Pressures of Some Ethers" J. Chem. Thermo. 8, 165 (1976).

2212 Alm, K., Ciprian, M., "Vapor Pressures, Refractive Index at 20 C., and Vapor-Liquid Equilibrium at 101.325 kPa in the Methyl tert-butyl ether-Methanol System," J. Chem. Eng. Data 25(2), 100 (1980).

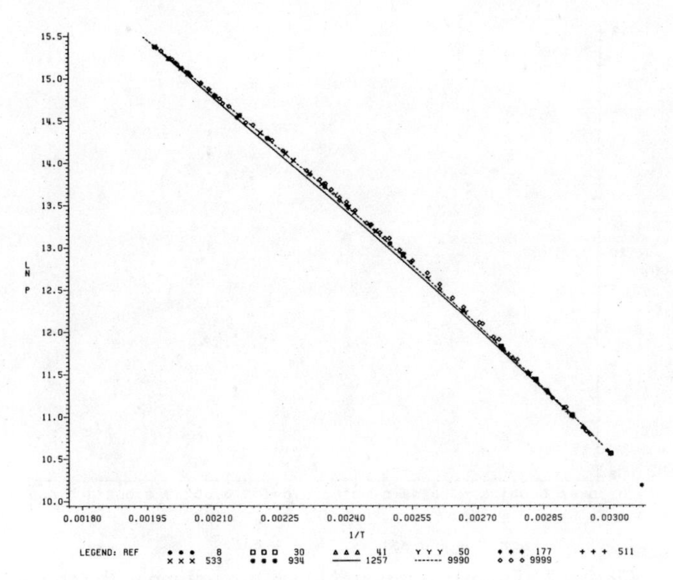

Figure 1. Experimental and literature values with regression line for isopropyl alcohol.

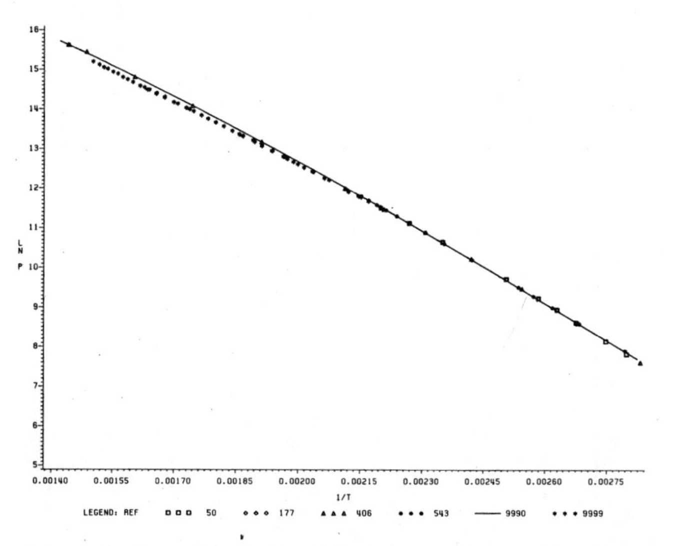

Figure 2. Experimental and literature values with regression line for phenol.

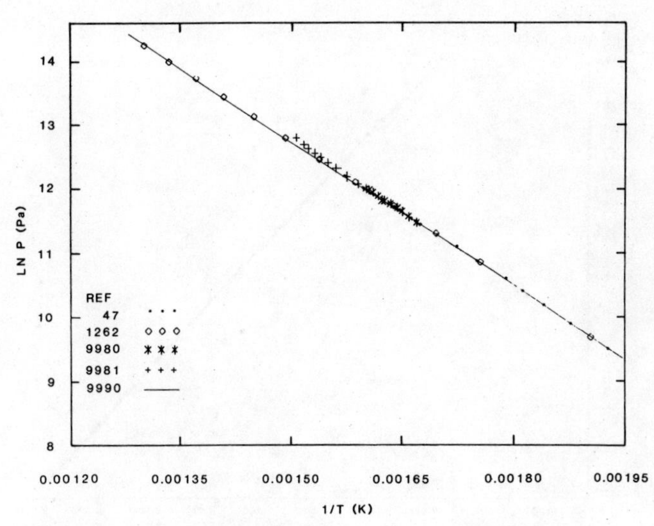

Figure 3. Experimental and literature values with regression line for *n*-dodecylbenzene.

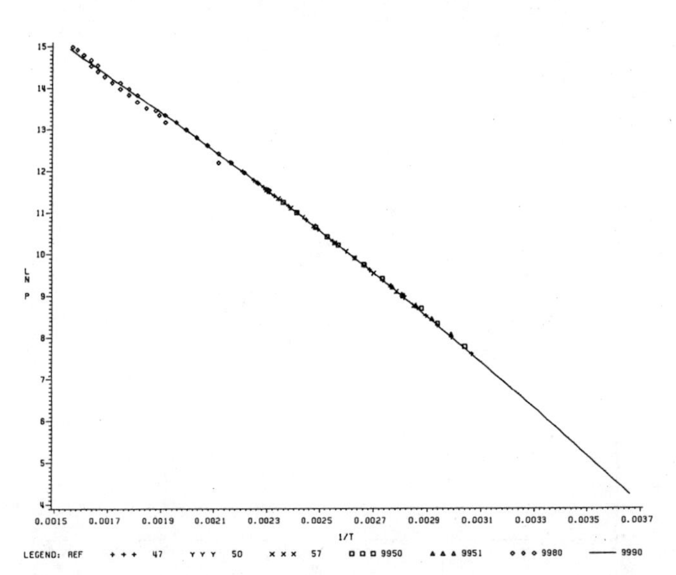

Figure 4. Experimental and literature values with regression line for *p*-ethyltoluene.

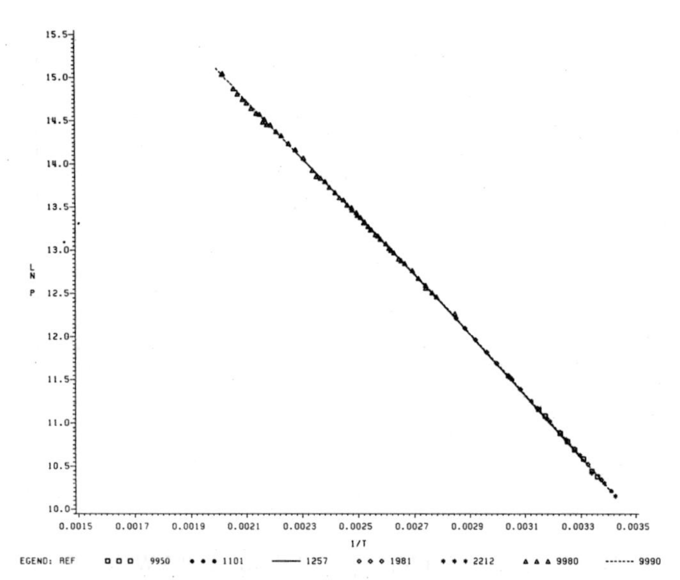

Figure 5. Experimental and literature values with regression line for methyl-*t*-butyl ether.

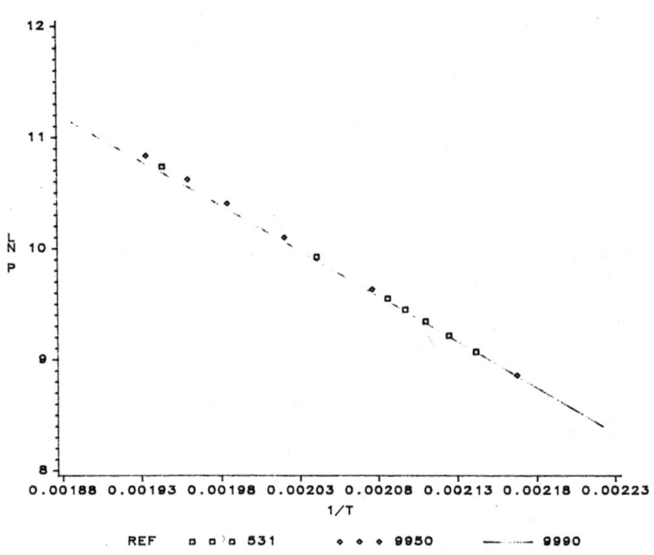

Figure 7. Experimental and literature values with regression line for diethanolamine.

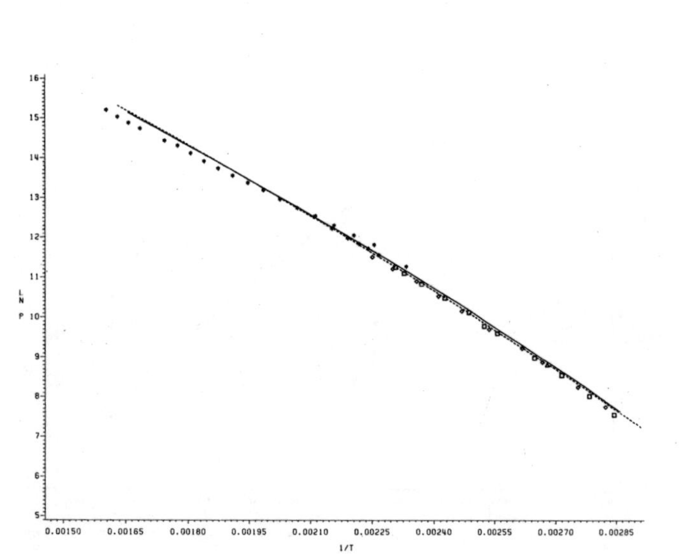

Figure 6. Experimental and literature values with regression line for monoethanolamine.

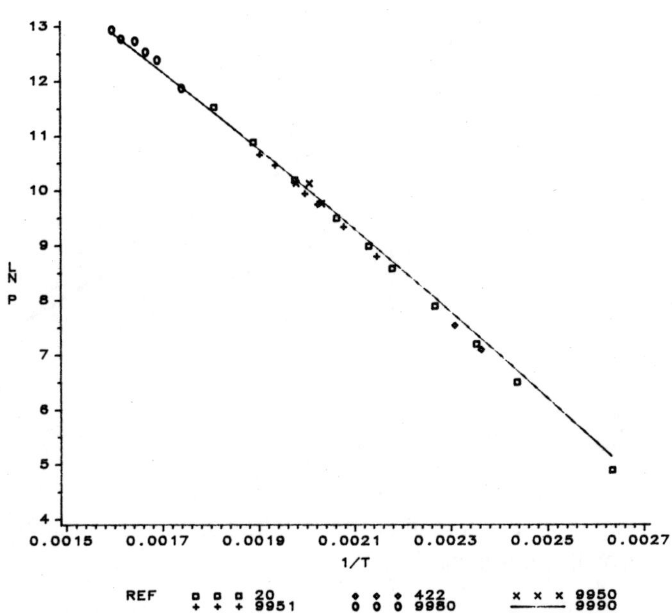

Figure 8. Experimental and literature values with regression line for 2,4-toluenediamine.

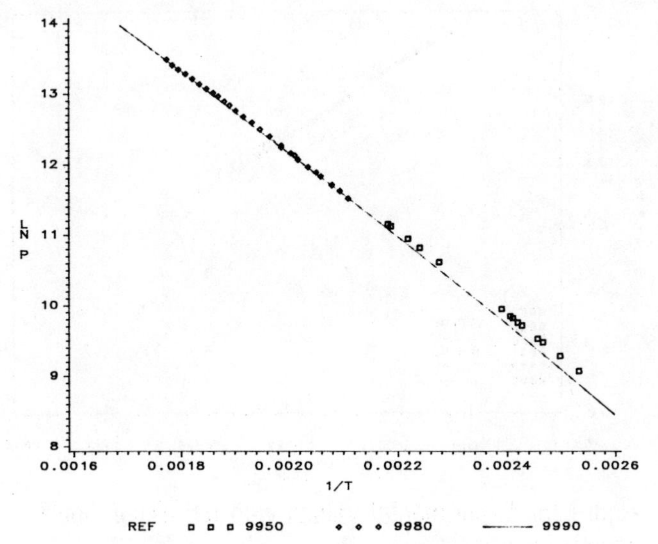

Figure 9. Experimental and literature values with regression line for hexamethylenediamine.

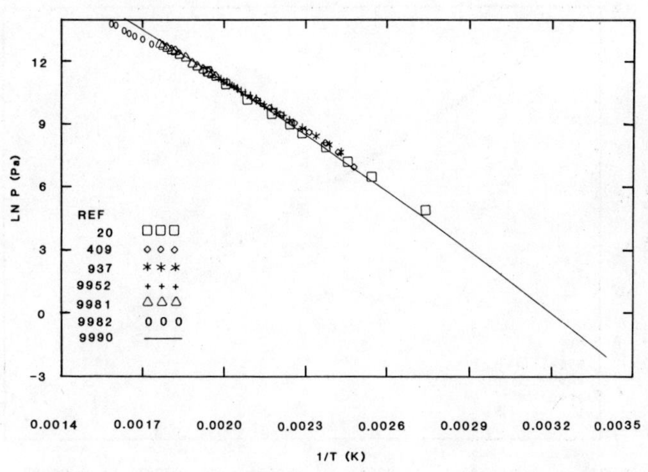

Figure 11. Experimental and literature values with regression line for diethylene glycol.

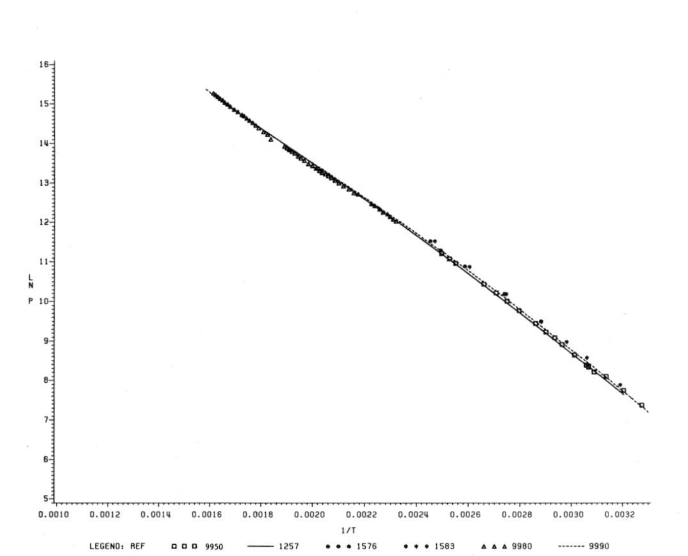

Figure 10. Experimental and literature values with regression line for hexamethyleneimine.

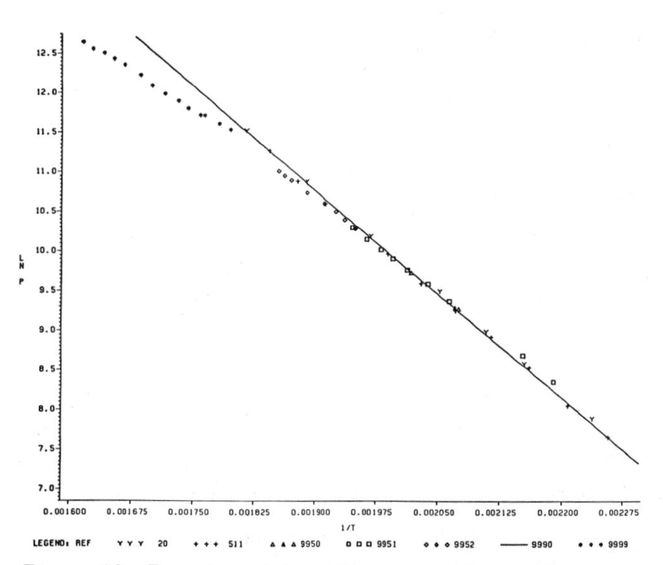

Figure 12. Experimental and literature values with regression line for triethylene glycol.

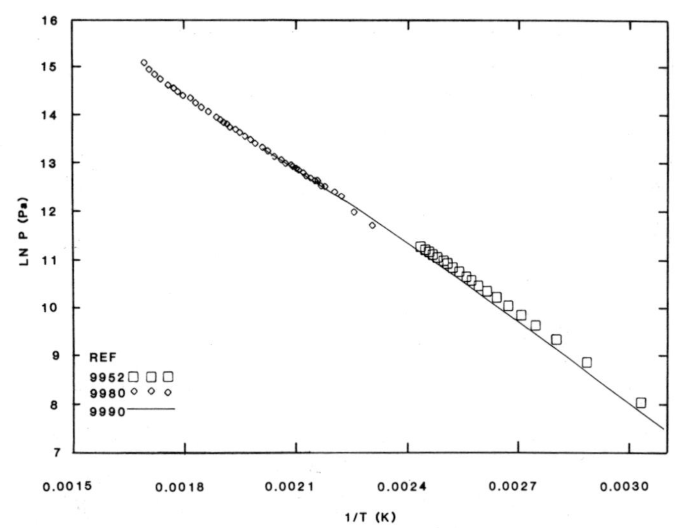

Figure 13. Experimental values with regression line for propylene glycol monomethyl ether acetate.

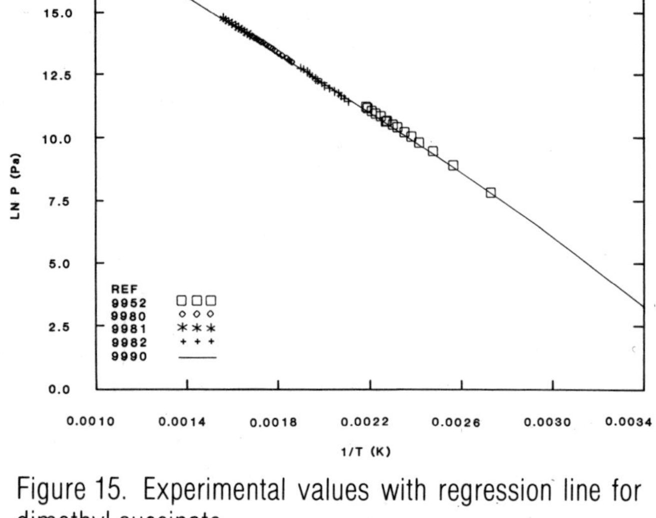

Figure 15. Experimental values with regression line for dimethyl succinate.

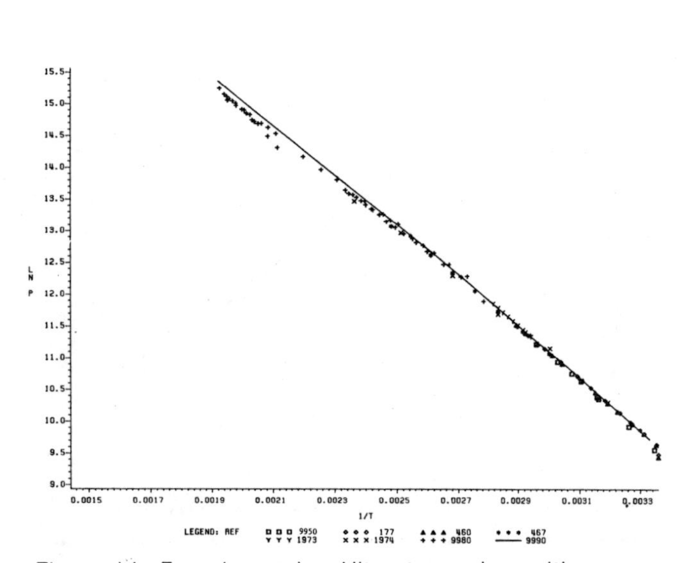

Figure 14. Experimental and literature values with regression line for vinyl acetate.

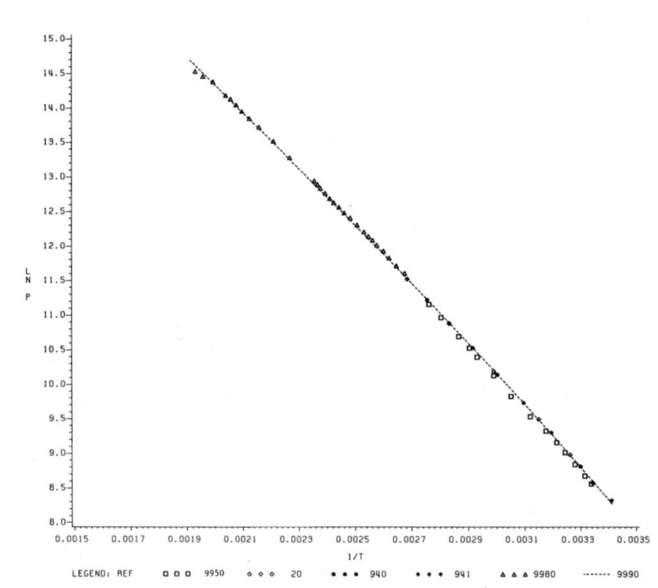

Figure 16. Experimental and literature values with regression line for ethyl acrylate.

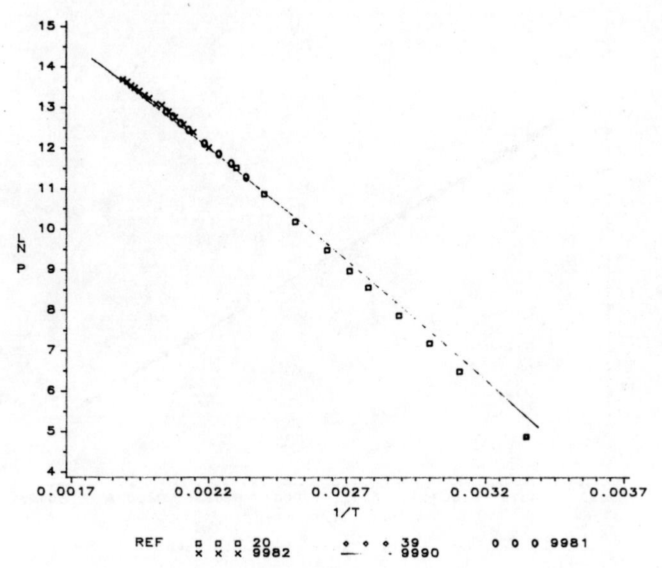

Figure 17. Experimental and literature values with regression line for methacrylic acid.

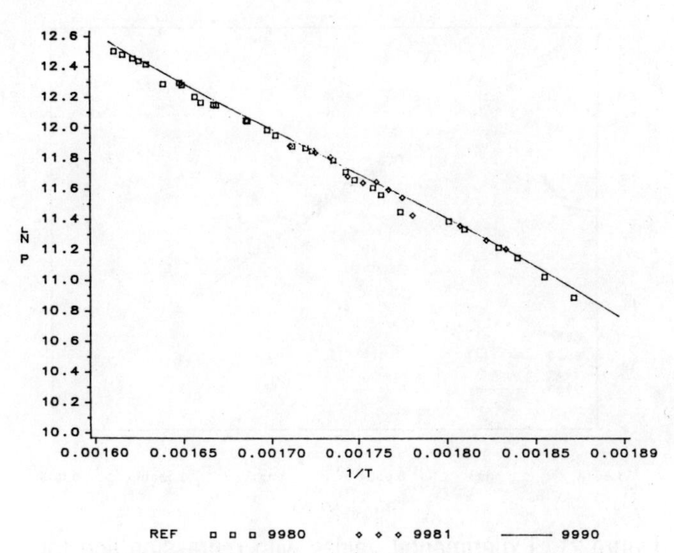

Figure 19. Experimental values with regression line for N-cyclohexylpyrrolidone.

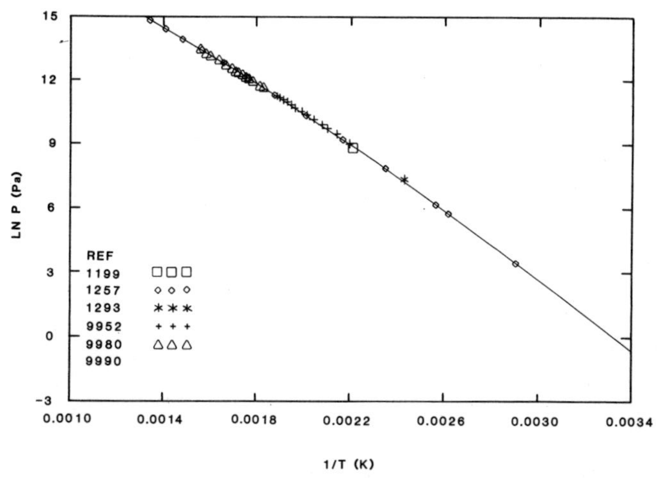

Figure 18. Experimental and literature values with regression line for ϵ-caprolactam.

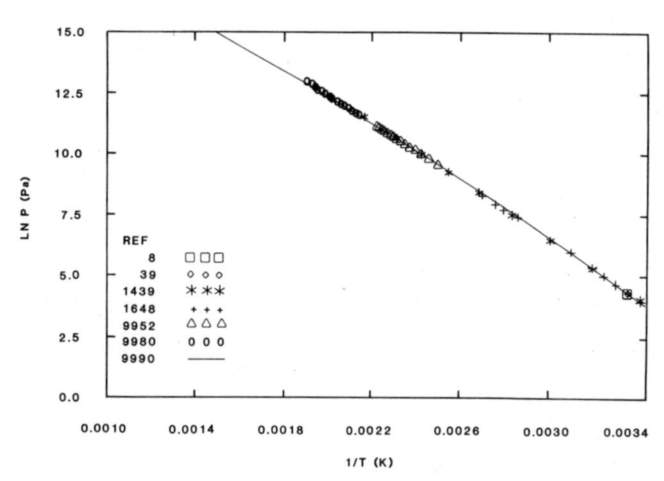

Figure 20. Experimental and literature values with regression line for dimethyl sulfoxide.

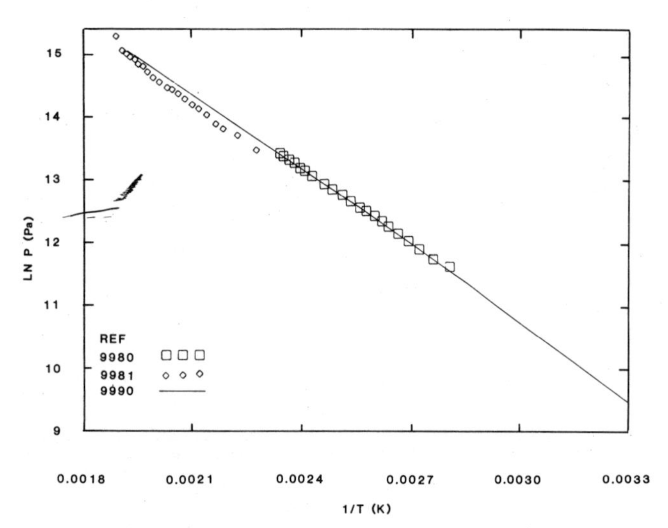

Figure 21. Experimental values with regression line for trimethoxysilane.

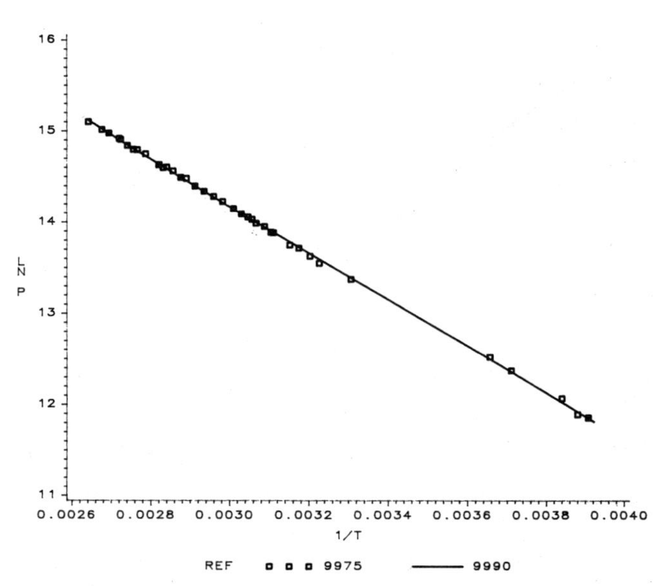

Figure 22. Experimenta values with regression line for 3,3,3,-trifluoropropene.

INDEX

SYMPOSIUM SERIES

ADSORPTION

96 Developments in Physical Adsorption
117 Adsorption Technology
219 Recent Advances in Adsorption and Ion Exchange

230 Adsorption and Ion Exchange—'83

233 Adsorption and Ion Exchange—Progress and Future Prospects
242 Adsorption and Ion Exchange: Recent Developments

AEROSPACE

33 Rocket and Missile Technology

52 Chemical Engineering Techniques in Aerospace

BIOENGINEERING

69 Bioengineering and Food Processing
84 The Artificial Kidney
86 Bioengineering ... Food
93 Engineering of Unconventional Protein Production

99 Mass Transfer in Biological Systems
108 Food and Bioengineering—Fundamental and Industrial Aspects
114 Advances in Bioengineering
163 Water Removal Processes: Drying and Concentration of Foods and Other Materials

172 Food, pharmaceutical and bioengineering—1976/77
181 Biochemical Engineering Renewable Sources of Energy and Chemical Feedstocks

CRYOGENICS

224 Cryogenic Processes and Equipment 1982

251 Cryogenic Properties, Processes and Applications 1986

CRYSTALLIZATION

110 Factors Influencing Size Distribution
193 Design Control and Analysis of Crystallization Processes

215 Nucleation, Growth and Impurity Effects in Crystallization Process Engineering
240 Advances in Crystallization From Solutions

253 Fundametnal Aspects of Crystallization and Precipitation Processes

DRAG REDUCTION

111 Drag Reduction

130 Drag Reduction in Polymer Solutions

ENERGY

Conversion and Transfer

5 Heat Transfer, Atlantic City
57 Heat Transfer, Boston
59 Heat Transfer, Cleveland
75 Energy Conversion Systems
79 Heat Transfer with Phase Change
87 Advances in Cryogenic Heat Transfer
113 Convective and Interfacial Heat Transfer

118 Heat Transfer—Tulsa
119 Commercial Power Generation
138 Heat Transfer—Research and Design
162 Energy and Resource Recover from Industrial and Municipal Solid Wastes
174 Heat Transfer, Research and Application
189 Heat Transfer—San Diego 1979

202 Transport with Chemical Reactions
208 Heat Transfer—Milwaukee 1981
216 Processing of Energy and Metallic Minerals
225 Heat Transfer—Seattle 1983
236 Heat Transfer—Niagra Falls 1984
245 Heat Transfer—Denver 1985
257 Heat Transfer—Pittsburgh 1987

Nuclear Engineering

53 Part XIII
56 Part XIV
94 Part XX
104 Part XXI

106 Part XXII
119 Commercial Power Generation
168 Heat Transfer in Thermonuclear Power Systems

169 Developments in Uranium Enrichment
191 Nuclear Engineering Questions Power Reprocessing, Waste, Decontamination Fusion
221 Recent Developments in Uranium Enrichment

ENVIRONMENT

78 Water Reuse
97 Water—1969
115 Important Chemical Reactions in Air Pollution Control
122 Chemical Engineering Applications of Solid Waste Treatment
124 Water—1971
126 Air Pollution and its Control
133 Forest Products and the Environment
137 Recent Advances in Air Pollution Control
139 Advances In Processing and Utilization of Forest Products
144 Water—1974: I. Industrial Wastewater Treatment
145 Water—1974: II. Municipal Wastewater Treatment
146 Forest Product Residuals
147 Air: I. Pollution Control and Clean Energy
148 Air: II. Control of NO_{xx} and SO_x Emissions
149 Trace Contaminants in the Environment
151 Water—1975
156 Air Pollution Control and Clean Energy
157 New Horizons for the Chemical Engineer in Pulp and Paper Technology

165 Dispersion and Control of Atmospheric Emissions, New-Energy-Source Pollution Potential
170 Intermaterials Competition in the Management of Shrinking Resources
171 What the Filterman Needs to Know About Filtration
175 Control and Dispersion of Air Pollutants: Emphasis on NO_X and Particulate Emissions
177 Energy and Environmental Concerns in the Forest Products Industry
184 Advances in the Utilization and Processing of Forest Products
188 Control of Emissions from Stationary Combustion Sources Pollutant Detection and Behavior in the Atmosphere
195 The Role of Chemical Engineering in Utilizing the Nation's Forest Resources
196 Implications of the Clean Air Amendments of 1977 and of Energy Considerations for Air Pollution Control
198 Fundamentals and Applications of Solar Energy
200 New Process Alternatives in the Forest Products Industries

201 Emission Control from Stationary Power Sources: Technical, Economic and Environmental Assessments
207 The Use and Processing of Renewable Resources—Chemical Engineering Challenge of the Future
209 Water—1980
210 Fundamentals and Applications of Solar Energy II
211 Research Trends in Air Pollution Control: Scrubbing, Hot Gas Clean-up, Sampling and Analysis
213 Three Mile Island Cleanup
223 Advances in Production of Forest Products
232 Applications of Chemical Engineering in the Forest Products Industry
239 The Impact of Energy and Environmental Concerns on Chemical Engineering in the Forest Products Industry
243 Separation of Heavy Metals and Other Trace Contaminants
246 Advances in Process Analysis and Development in the Forest Products Industries.

FLUIDIZATION

01 Fundamental Processes in Fluidized Beds
05 Fluidization Fundamentals and Application
16 Fluidization: Fundamental Studies Solid-Fluid Reactions, and Applications

176 Fluidization Application to Coal Conversion Processes
205 Recent Advances in Fluidization and Fluid-Particle Systems

234 Fluidization and Fluid Particle Systems: Theories and Applications
241 Fluidization and Fluid Particle Systems: Recent Advances
255 New Developments in Fluidization and Fluid-Particle Systems

HISTORY OF CHEMICAL ENGINEERING

100 The History of Penicillin Production
235 Diamond Jubilee Historical/Review Volume

ION EXCHANGE

179 Adsorption and Ion Exchange Separations
219 Recent Advances in Adsorption and Ion Exchange
230 Adsorption and Ion Exchange—'83
233 Adsorption and Ion Exchange—Progress and Future Prospects

KINETICS

25 Reaction Kinetics and Unit Operations
73 Kinetics and Catalysis

MINERALS

15 Mineral Engineering Techniques
85 Fossil Hydrocarbon and Mineral Processing
173 Fundamental Aspects of Hydrometallurgical Processes
180 Spinning Wire from Molten Metals
216 Processing of Energy and Metallic Minerals

PETROCHEMICALS

49 Polymer Processing
127 Declining Domestic Reserves—Effect on Petroleum and Petrochemical Industry
135 The Petroleum/Petrochemical Industry and the Ecological Challenge
142 Optimum Use of World Petroleum
212 Interfacial Phenomena in Enhanced Oil Recovery

PETROLEUM PROCESSING

103 C_4 Hydrocarbon Production and Distribution
127 Declining Domestic Reserves—Effect on Petroleum and Petrochemical Industry
135 The Petroleum/Petrochemical Industry and the Ecological Challenge
142 Optimum Use of World Petroleum
155 Oil Shale and Tar Sands
226 Underground Coal Gasification: The State of the Art

PHASE EQUILIBRIA

2 Pittsburgh and Houston
6 Collected Research Papers
88 Phase Equilibria and Gas Mixtures Properties

PROCESS DYNAMICS

36 Process Dynamics and Control
46 Process Systems Engineering
55 Process Control and Applied Mathematics
214 Selected Topics on Computer-Aided Process Design and Analysis

SEPARATION

120 Recent Advances in Separation Techniques
192 Recent Advances in Separation Techniques—II
250 Recent Advances in Separation Techniques—III

SONICS

109 Sonochemical Engineering

MISCELLANEOUS

48 Chemical Engineering Reviews
70 Small-Scale Equipment for Chemical Engineering Laboratories
112 Engineering, Chemistry, and Use of Plasma Reactors
125 Vacuum Technology at Low Temperatures
143 Standardization of Catalyst Test Methods
182 Biorheology
183 The Modern Undergraduate Laboratory Innovative Techniques
185 Electro Organic Synthesis Technology
186 Plasma Chemical Processing
187 Chronic Replacement of Kidney Function
194 Hazardous Chemical—Spills and Waterborne Transportation
203 A Review of AIChE's Design Institute for Physical Property Data (DIPPR) and Worldwide Affiliated Activities
204 Tutorial Lectures in Electrochemical Engineering and Technology
206 Controlled Release Systems
217 New Composite Materials and Technology
220 Uncertainty Analysis for Engineers
228 Problem Solving
229 Tutorial Lectures in Electrochemical Engineering and Technology—II
231 Data Base Implementation and Application
237 Awareness of Information Sources
238 New Developments in Liquid-Liquid Extractors: Selected Papers From ISEC '83
244 Experimental Results from the Design Institute for Physical Property Data. I: Phase Equilibria
247 Chemical Engineering Data Sources
248 Industrial Membrane Processes
249 Measurement of High Temperatures in Furnaces and Processes
252 Thin Liquid Film Phenomena
254 Electrochemical Engineering Applications
256 Experimental Results From the Design Institute for Physical Property Data: Phase Equilibria and Pure Component Properties
258 Fiber Optics: Processing and Applications

MONOGRAPH SERIES

3 The Manufacture of Nitric Acid by the Oxidation of Ammonia—The DuPont Pressure Process by Thomas H. Chilton
4 Experiences and Experiments with Process Dynamics by Joel O. Hougen
5 Present, Past, and Future Property Estimation Techniques by Robert C. Reid
6 Catalysts and Reactors by James Wei
7 The 'Calculated' Loss-of-Coolant Accident by L.J. Ybarrondo, C.W. Solbrig, H.S. Isbin
8 Understanding and Conceiving Chemical Process by C. Judson King
9 Ecosystem Technology: Theory and Practice by Aaron J. Teller
10 Fundamentals of Fire and Explosion by Daniel R. Stull
11 Lumps, Models and Kinetics in Practice by Vern W. Weekman, Jr.
12 Lectures in Atmospheric Chemistry by John H. Seinfeld
13 Advanced Process Engineering by James R. Fair
14 Synfuels from Coal by Bernard S. Lee
15 Computer Modeling of Chemical Processes by J.D. Seader
16 "High-Tech" Materials by Sheldon Isakoff